D1429864

Howard M. Berlin

EXPERIMENTS IN ELECTRONIC DEVICES

Fourth Edition

To Accompany
FLOYD'S
ELECTRONIC DEVICES
and
ELECTRONIC DEVICES:
ELECTRON-FLOW VERSION

310601

Prentice Hall
Englewood Cliffs, New Jersey   Columbus, Ohio

Cover photo: Copyright © Superstock
Editors: Dave Garza and Judith Casillo
Developmental Editor: Carol Hinklin Robison
Production Editor: Rex Davidson
Cover Designer: Brian Deep
Production Manager: Patricia A. Tonneman
Marketing Manager: Debbie Yarnell

This book was printed and bound by Quebecor Printing/Book Press. The cover was printed by Phoenix Color Corp.

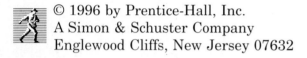

© 1996 by Prentice-Hall, Inc.
A Simon & Schuster Company
Englewood Cliffs, New Jersey 07632

Printed in the United States of America

10 9 8 7 6 5 4 3 2 1

ISBN: 0-13-399544-5

Prentice-Hall International (UK) Limited, *London*
Prentice Hall of Australia Pty. Limited, *Sydney*
Prentice-Hall Canada Inc., *Toronto*
Prentice-Hall Hispanoamericana, S. A., *Mexico*
Prentice-Hall of India Private Limited, *New Delhi*
Prentice-Hall of Japan, Inc., *Tokyo*
Simon & Schuster Asia Pte. Ltd., *Singapore*
Editora Prentice-Hall do Brasil, Ltda., *Rio de Janeiro*

PREFACE

This laboratory workbook is designed for use with the texts *Electronic Devices,* Fourth Edition, and *Electronic Devices: Electron-Flow Version* Second Edition by Thomas L. Floyd. The 43 experiments cover virtually every basic aspect of circuits containing rectifier and zener diodes; bipolar, field effect (JFET, MOSFET), and unijunction transistors; silicon-controlled rectifiers; operational amplifiers; and integrated-circuit voltage regulators, timers, and phase-locked loops.

Although the experiments are specifically referenced to the companion text, they are nevertheless general enough to be easily integrated with any other textbook on semiconductor devices at the electrical/electronics technology level. The experiments in this book reinforce and expand upon the concepts presented in the classroom. The student is able to verify these concepts by performing detailed step-by-step experiments that are easily accomplished in a typical two- to three-hour lab session. In all cases, experimental measurements can be reasonably compared to theory. Although there are 43 experiments, the student will be expected to perform only a select number of them. It is not necessary to treat all the experiments separately; some may be conveniently combined as a single major experiment. Experiments 9 through 12, for example, concentrate on the biasing of bipolar transistors and can easily be consolidated.

Several features enhance the utility of this workbook:

1. The purpose of each experiment is clearly defined. In addition, a short background summary on the operation of the circuit to be investigated is included.
2. At selected portions in a number of experiments, photographs of the oscilloscope's display give students a picture of what they should observe on the screen.
3. For each experiment, a list of required parts and test equipment is included. All parts are low in cost and are readily obtainable from a number of sources, including Radio Shack. For convenience, a summary list of these required parts appears in the Appendix.
4. When appropriate, a summary of useful formulas is included to enable the student to compare measured results to theory.
5. A summary of "What You Have Done," presented at the end of each experiment, restates and re-emphasizes the main points of the experiment as stated in the "Purpose and Background" section of the experiment.
6. There are "student response" pages at the end of each experiment. These pages are for the student to enter the objectives/purpose, the schematic diagram(s), all measured data (on a blank graph page when required), and the results and conclusions of the experiment. These pages are perforated so that they, along with the optional review questions, can be easily removed to be submitted to the instructor as a complete laboratory report.
7. Multiple-choice review questions are included.

I would like to thank the many users of the previous editions who have offered their helpful suggestions for improvement. I also wish to express my gratitude to the reviewers, who offered both praise and constructive suggestions: Richard Burchell, Riverside City College; Ronald Emery, Indiana University–Purdue University; Gary House, DeVry Institute of Technology—Atlanta; Maurice Nadeau, Central Minnesota Vocation-Technical Institute; Tim Staley, DeVry Institute of Technology—Dallas; Guy Tolbert, Surry Community College; Ulrich Zeisler, Utah Technical College at Salt Lake City, Steve Harsanny, Mount San Antonio College, who checked this manual for accuracy; and especially Thomas L. Floyd, the author of the text to which this book is a companion.

Howard M. Berlin

CONTENTS

PERFORMING THE EXPERIMENTS

EXPERIMENTS

APPENDIX

PERFORMING THE EXPERIMENTS

INTRODUCTION

A laboratory experiment, although a powerful learning tool in the educational process, is a double-edged sword. In order to receive the benefits it can provide, you must follow several rules so that your experiment will be successful. This section illustrates these rules and describes how each of the 43 experiments is presented.

BREADBOARDING

The breadboard is designed to accommodate the experiments that you will perform. The various transistors, diodes, integrated circuit devices, resistors, capacitors, and other components, as well as power and signal connections, all tie directly to the breadboard. Figure 1 shows the top view of the "solderless" breadboarding socket, which is manufactured or sold by several companies, including AP Products, Continental Specialties, and Radio Shack.

Breadboarding is an art that cannot be learned in a few minutes. Rather, it takes practice and experience to develop an efficient technique. An artist plans his creation, making sure that the picture

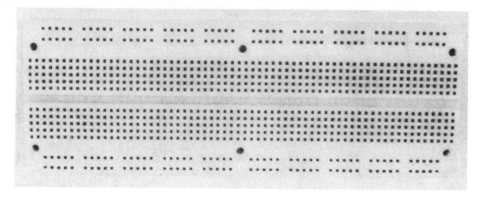

FIGURE 1

will fit on the canvas in the proper proportions without crowding, and the same is true for breadboarding electronic circuits.

When breadboarding, keep the following rules in mind:

1. Only no. 22, 24, or 26 insulated wire should be used, and it must be *solid*, not stranded!
2. Never insert too large a wire or component lead into a breadboarding terminal.
3. Never insert a bent wire. Straighten out the bent end with a pair of pliers before insertion.
4. Try to maintain an orderly arrangement of components and wires, keeping all connections as short as possible. Generally, the circuit is arranged on the breadboard in the same way that it appears on a schematic diagram. This rule is useful when you are trying to locate possible wiring errors.

RULES FOR SETTING UP THE EXPERIMENTS

Throughout this laboratory workbook, you will have the opportunity to breadboard a variety of circuits. Before setting up any experiment, you should do the following:

1. Plan your experiment. Know what types of results you are expected to observe.
2. Disconnect, or turn off, *all* power and external signal sources from the breadboard.
3. Clear the breadboard of all wires and components from previous experiments, unless instructed otherwise.
4. Check the wired-up circuit against the schematic diagram to make sure that it is correct.

5. Unless otherwise instructed, never make component or wiring changes on or to the breadboard with the power or external signal connections to the breadboard. This rule reduces the possibility of accidentally destroying electronic components and equipment.

6. When you have finished, make sure that you disconnect everything *before* you clear the breadboard of wires and components.

FORMAT FOR THE EXPERIMENTS

The instructions for each experiment are presented in the following format:

1. **Purpose and Background.** The material under this heading states the purpose of the experiment. You should have this purpose in mind as you conduct the experiment. In addition, there is a short summary about the operation and characteristics of the circuit that you will be building.

2. **Text Reference.** The corresponding section number and title in *Electronic Devices* are given here. These sections discuss the background for the experiment.

3. **Required Parts and Equipment.** A listing of the required circuit components and test equipment necessary for the experiment is given under this heading. Virtually all parts are low-cost and readily available from a variety of commercial sources, including local Radio Shack stores. (A list of the necessary components needed for *all* the experiments is given in the Appendix; in most cases, each component also includes Radio Shack's catalog number.)

    Several pieces of equipment will be required for the experiments:

    Oscilloscope. Just about any general-purpose type will do, and it should be a dual trace type. Input sensitivity generally ranges from 5 mV/division to 10 V/division and has a bandwidth from 5 MHz to 20 MHz. When needed, the schematic symbol of Figure 2 will be used.

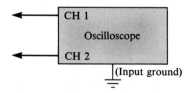

FIGURE 2

VOM, VTVM, or DMM. A general-purpose type capable of measuring dc and ac voltages and current, as well as resistance, is necessary. If you can obtain one, use a digital type; otherwise, any VOM used should have at least a 50 kΩ/V rating so as not to introduce serious loading errors.

Adjustable dc power supply. The dc power supply should have an adjustable output voltage range from 0 V to at least 15 V, with the capability of delivering 500 mA. Some experiments will require two power supplies.

Frequency counter. It need not be an expensive one, but it should have a resolution of 1 Hz for precise measurements. There are several units available in kit form for less than $80.

Function generator. A function generator is a signal source capable of producing selectable sine, triangle, and square waveforms of variable frequency and amplitude. Generally, they will have voltage levels from several millivolts to approximately 20 V peak-to-peak with an output impedance to 50 Ω over a frequency range from 1 Hz to 100 kHz. Also, there is usually a TTL level pulse output. When needed, the schematic symbol of Figure 3 will be used.

Sine wave   Square wave

FIGURE 3

4. **Useful Formulas.** Under this heading is a summary of the equations, when applicable, that apply to the design and operation of the circuit. These formulas are presented so that you can compare measured results with theory.

5. **Procedure.** First the following diagrams are presented:

Schematic diagram of circuit. This figure shows the completed circuit that you will wire up for the experiment. You should analyze this diagram in an effort to obtain an understanding of the circuit *before* you proceed. When used, the oscilloscope connections to the circuit are shown with bolder lines so that they will not be confused with the normal circuit.

Pin configuration. These configurations are given for all transistors and integrated circuit devices used in the experiment.

Next, a series of numbered, sequential steps gives detailed instructions for performing portions of the experiment. When appropriate, photographs of the oscilloscope's display are included so that you will be able to compare your results. In addition, questions are included at appropriate points. Any numerical calculations are performed easily on many of the pocket-type calculators.

6. **What You Have Done.**   Under this heading is an explanation of the main points gained in performing the experiment.

HELPFUL HINTS AND SUGGESTIONS

Besides the necessary parts and equipment, only three small hand tools are necessary for all of the experiments given in this book:

1. A pair of long-nosed pliers
2. A wire stripper/cutter
3. A small screwdriver

The *pliers* are used to straighten out the bent ends of hookup wires that are used to wire the circuits on the breadboard. They are also used to straighten out or bend the resistor, capacitor, and other component leads to the proper position so that they can be conveniently inserted into the breadboard.

The *wire stripper/cutter* is used to cut the hookup wire to size and strip about $3/8$ inch of insulation from each end.

The *screwdriver* can be used to remove the integrated-circuit devices from the solderless breadboarding socket by gently prying them loose.

In general, each experiment will ask you to compare a measured value with an expected value, that is, a value you would expect to obtain if you had not done the experiment. Generally, this value will be determined from theory. To compare your experimental result with the predicted value, calculate the *percent error*, or *percent difference*. The formula for determining the percent error is

$$\% \text{ error } = \frac{\text{measured value} - \text{true value}}{\text{true value}} \times 100\%$$

For example, if you measure a voltage of 5.36 V and you expect the value (from theory) to be 4.97 V, the percent error is

$$\% \text{ error } = \frac{5.36 - 4.97}{4.97} \times 100\% = 7.8\%$$

The most common sources of error are the tolerance of component values and the loading effects of meters. The measured result is generally acceptable if the percent error is within 10%.

THE LABORATORY REPORT

As with any laboratory-oriented course, there is usually more to the laboratory exercise than merely performing the experiment. Performing the laboratory experiment should be followed by writing what was done in a technical report. As an educational exercise, writing the report serves two major purposes. First, a written report documents what took place in the laboratory and gives the instructor an indication of how well the student has understood the principles and concepts that were supposed to be demonstrated by a particular experiment. Second, writing a *technically* oriented report provides the student with the practice of developing the communications skills that will be useful in his or her professional career. It is my experience that many students, even in standard courses in English composition, receive little or no instruction in writing technical material about their particular area of study.

Many instructors or departments may have their own guidelines and rules concerning the format of the laboratory report. Following is a brief discussion of the guidelines used in the Electrical and Electronics Technology program here at Delaware Technical and Community College, Stanton Campus.

The completed laboratory report may be either neatly handwritten in *ink* (black or blue) or typewritten. The point here is that *neatness* and *legibility* are important. If the instructor cannot read a sloppily written or typed report, then the effort and time taken in its preparation has been wasted. In our courses, ample time (usually one week) is given from the date the experiment was performed until the report is due. A report should contain the following elements:

1. **Title.** The title of the experiment at the beginning of the report should be a descriptive phrase that identifies the experiment. As an example, "The Common-Emitter Amplifier" is more descriptive than "Experiment 4" because it may not be apparent what the substance of Experiment 4 actually is. In addition, the name of the person submitting the report should be included, along with the names of other laboratory partners. Finally, the date the experiment was performed must be included.

2. **Purpose or Objective.** The particular experiment is to be performed for some definite reason. A brief statement (one to three sentences) that explains why the experiment is being performed is included.

3. **Wiring (Schematic) Diagram.** A schematic diagram of the circuit(s) used in the experiment must be included. The diagram must be neatly drawn and properly labeled with all

component values used. The use of templates to draw electronic symbols is not mandatory, and the symbols can be drawn freehand. However, all connecting lines must be drawn with a straightedge.

4. **Equipment and Special Supplies.** The serial number and model number of the equipment used in the experiment should be recorded. In some cases, the student may be asked to repeat the experiment to show why unusual results have been obtained. In practice, faulty results may be obtained because of a faulty meter or other equipment, a fact the student can prove *only* if the exact instrument is used in the repeated experiment.

5. **Procedure.** This contains short comments in chronological order about the measurements, instruments used, and any special techniques used. Such comments are usually sufficient to explain what was done during the experiment. Long explanations are generally not necessary and usually not desirable. The comments nevertheless should be complete enough to allow another person to perform the experiment for verification.

6. **Data Tables.** In almost all experiments, a number of measured values should be recorded *in ink*. Charts are the most convenient method of recording these data, since all values are readily available for analysis. The data sheet on the perforated page at the end of each experiment in this laboratory workbook serves as the format for the data table.

   The student should also understand that all erasures and changes in data are observed by others with great suspicion. Recording errors should not be erased. They should be indicated as errors by drawing a *single* line through the incorrect data and writing the correct value beside it. Many industrial firms also require that the employee initial the change along with the date. Indicating the error by the single-line method retains the original measurement for future analysis if it should be found to be important. In some experiments, the possible error is the only interesting part of the experiment, and the measurement would be completely lost if it were erased, scratched out, or otherwise defaced.

7. **Calculations.** All experiments require a certain number of calculations before final results are obtained. Sample calculations that are completely identified should be included. It is not necessary to show repeated calculations.

8. **Graphs.** Data that have been recorded in long columns on charts cannot be analyzed quickly, but the graph provides a visual, or pictorial, presentation of the data. The graph can be used to determine trends and unusual results, such as a data point that is probably in error.

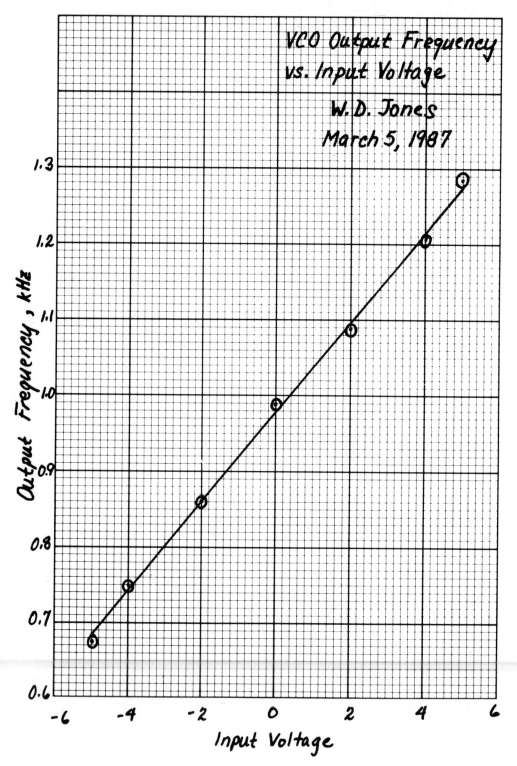

FIGURE 4 *Sample hand-drawn graph.*

The guidelines in constructing a graph are very simple:

- Always label the axes. Generally, one parameter is plotted as another parameter is varied. The *dependent variable* is represented on the vertical axis, and the *independent variable* on the horizontal axis. If, for example, the gain of the amplifier is measured as the input frequency is changed, the amplifier gain is the dependent variable because its value depends on the value of the input frequency.
- All graphs should include a title, name of the preparer, and the date the graph was made. At some later time, this basic information pins down what the graph depicts, who did it, and when it was done.
- If various conditions plotted on the same graph result in more than one curve or line, the data points from one set of data should be easily distinguishable from data points belonging to a different set. Data points are frequently marked using a dot surrounded by a small circle, or Δ, $+$, $\times$, $*$ symbols.
- The data points should be connected smoothly with the best-fit straight line between all points, or with a "French curve" to construct a smooth curve between points. Figure 4 shows an example of a completed hand-drawn graph of the measured output frequency of an oscillator as the input voltage is varied.

9. **Results and Conclusions.** The results and conclusions are probably the most important part of the experiment. The entire experiment is considered a failure if the student does not understand the results and cannot decide how to express the conclusion. Many instructors read this part of the report first and then refer to the first eight items for supporting information. The conclusion should be as brief as possible (less than one written page). Long conclusions often tend to bury the actual results of the experiment and instead become a procedure sheet. One statement that should be avoided is "Everything went well, as expected."

10. **Answers to Review Questions** (optional). At the end of each experiment in the workbook are several multiple-choice questions. Some instructors may want to have the student include the answers to these questions as part of the laboratory report.

1

THE DIODE

PURPOSE AND BACKGROUND

The purpose of this experiment is to examine characteristics of a silicon diode. When the diode's anode is at a higher potential than is the cathode, the diode is *forward biased*. For **conventional current flow**, current will flow through the diode from anode to cathode. For **electron flow**, current will flow from cathode to anode. Unlike a resistor, in which the current is directly (that is, linearly) proportional to the voltage across it, the diode is a *nonlinear* device. When the diode is forward biased, a small but measurable voltage drop, called the *barrier potential*, occurs across the diode. For germanium diodes, this value is typically 0.3 V; for silicon diodes, it is approximately 0.7 V.

    Text Reference: 1–9, The Diode

REQUIRED PARTS AND EQUIPMENT

Resistors (1/4 W):
- ☐ 10 Ω
- ☐ 100 Ω
- ☐ 1 kΩ
- ☐ 1N914 (1N4148 or equivalent) silicon, small-signal diode

- ☐ 0–15 V dc power supply
- ☐ Signal generator
- ☐ Dual trace oscilloscope
- ☐ VOM
- ☐ Breadboarding socket

11

USEFUL FORMULA

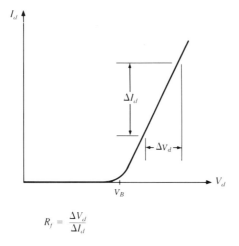

$$R_f = \frac{\Delta V_d}{\Delta I_d}$$

FIGURE 1–1   *Graphic determination of diode forward resistance.*

The determination of R_f, the diode forward resistance, is shown graphically in Figure 1–1.

$$R_f = \frac{\Delta V_d}{\Delta I_d}$$

PROCEDURE

1. Very often one can use a VOM to check quickly whether a diode is good or bad. Unless they have a specific function for this purpose, most DMMs are not able to perform this test properly. Using your VOM as an *ohmmeter*, first select a low-resistance meter range, such as the "R × 100" range. Then connect the *positive* lead of the VOM to the diode's *anode* terminal while the *negative* lead is connected to the diode's *cathode* terminal, as shown in Figure 1–2A. (Most diodes have a single colored band, several bands, the diode symbol, or the letter "K" at one end to indicate the cathode terminal.) The VOM's internal battery then *forward biases the diode.* Note the resistance reading.

   If a DMM with a " diode check" feature is used, the display usually indicates the voltage drop across a good diode from anode to cathode when it is forward biased. When reverse biased, the DMM generally indicates some form of out-of-range condition, such as a blinking display or the letters "OL."

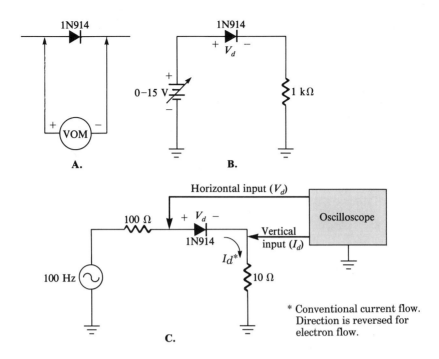

FIGURE 1–2 *Schematic diagram of circuits.*

2. Now reverse the VOM's leads so that the meter's positive terminal is connected to the cathode terminal of the diode, which is now reverse biased. Note the resistance reading.

The reading you have just obtained should be much *higher* (typically several hundred thousand ohms) compared to the resistance reading of Step 1, which is typically a few hundred ohms or less. Consequently, the diode exhibits a low *forward* resistance while having a high, or nearly infinite, *reverse* resistance. The actual resistance readings obtained are not as important as their relationship to each other. If both readings indicate virtually the same low resistance, then the diode is shorted; if a very high resistance is obtained in both directions, the diode is open.

When measuring resistances, some VOMs have the polarity of their leads reversed from the normal sense. That is, the positive lead is actually wired to the internal battery's *negative* terminal. In this case, the forward and reverse resistance readings will be the opposite of those indicated in these two steps. When it functions as a voltmeter or an ammeter, this type of VOM has its leads internally connected in the normal sense.

3. Wire the circuit shown in Figure 1–2B. Adjust the dc power supply to give the voltages across the 1-kΩ resistor shown in

Table 1–1. For each voltage, measure and record the dc voltage drop (V_d) across the diode. The diode current is also the current flowing through the 1-kΩ resistor. Determine the diode current by using Ohm's law in each case.

4. Plot the resulting diode curve (diode current versus voltage) on the graph page in this experiment. Graphically determine the diode's barrier potential (V_B) and forward resistance (R_F), recording your results in Table 1–2.

5. Disconnect the power from the breadboard and wire the circuit shown in Figure 1–2C. In this part, the oscilloscope is set up to function as an X-Y plotter. Set the oscilloscope controls to the following approximate settings:

> Vertical (or Y) input sensitivity: 10 mV/division, dc coupling

> Horizontal (or X) input sensitivity: 1 V/division, dc coupling

6. After the oscilloscope has warmed up, center the trace dot at the center of the scope's screen. Adjust the sine wave frequency of the signal generator to approximately 100 Hz, and vary the generator's output level so that you observe the characteristic diode curve similar to the one plotted in Step 4. The oscilloscope display should be similar to that shown in Figure 1–3. If it is not, the leads of the oscilloscope may be interchanged or there may be wiring error.

The horizontal input measures the voltage across the diode (V_d), neglecting the small voltage drop across the 10-Ω resistor. The vertical input measures the voltage drop across the 10-Ω

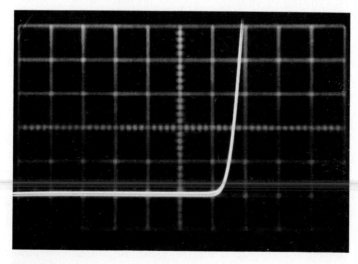

FIGURE 1–3

resistor. By Ohm's law, the vertical input can be made to show the diode current (I_d). If the vertical sensitivity is 10 mV/division, then in terms of the current through the 10-Ω resistor, which is the same as the diode current,

$$\text{Vertical sensitivity} = \frac{10 \text{ mV/division}}{10 \ \Omega}$$

$$= 1 \text{ mA/division}$$

7. As in Step 4, from the oscilloscope's display graphically determine the diode's barrier potential and forward resistance, recording your results in Table 1–2. How does this compare with Step 4 for the same diode?

WHAT YOU HAVE DONE

This experiment examined the characteristics of a silicon diode. You learned how to properly test a diode using a VOM or DMM, and how to determine the diode's barrier potential and forward resistance. By graphing the diode's forward characteristic, you observed that the diode is a nonlinear device.

NOTES

Name _____ Date _____

THE DIODE

OBJECTIVES/PURPOSE:

SCHEMATIC DIAGRAM:

Name _____ Date _____

DATA FOR EXPERIMENT 1

TABLE 1–1

Voltage across 1-kΩ Resistor	Diode Voltage	Diode Forward Current
0.1 V		
0.2 V		
0.3 V		
0.4 V		
0.5 V		
0.6 V		
0.7 V		
0.8 V		
0.9 V		
1 V		
2 V		
3 V		
4 V		
5 V		
6 V		
7 V		
0 V		
9 V		
10 V		

Name _____ Date _____

TABLE 1–2

Parameter	Step 4	Step 7
Diode barrier potential, V_B		
Diode forward resistance, R_F		

NOTES

Name _____ Date _____

DATA FOR EXPERIMENT 1

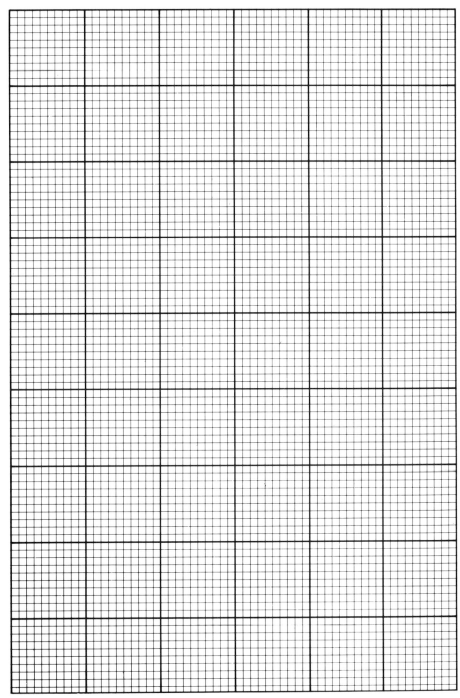

Name _____ Date _____

RESULTS AND CONCLUSIONS:

REVIEW QUESTIONS FOR EXPERIMENT 1

1. When an ohmmeter is used to test a diode, as in Figure 1–2A, a very low resistance (but not zero) in one direction means that the diode is
   (a) open   (b) shorted
   (c) forward biased   (d) reverse biased   (   )
2. In this experiment, the measured diode barrier potential is approximately
   (a) 0.3 V   (b) 0.6 V   (c) 0.9 V   (d) 1.2 V   (   )
3. If the 10-Ω resistor in Figure 1–2C is changed to 100 Ω and the oscilloscope's vertical sensitivity is 0.5 V/division, then the vertical axis, in terms of current, is
   (a) 0.5 mA/division   (b) 5 mA/division
   (c) 50 mA/division   (d) 0.5 A/division   (   )
4. For which region of your experimental diode curve does the diode look like an open circuit?
   (a) Diode voltages less than the barrier potential
   (b) Diode voltages greater than the barrier potential   (   )
5. For the region of the diode curve greater than the diode's barrier potential,
   (a) the curve is essentially horizontal
   (b) the diode forward resistance approaches an open circuit
   (c) the diode voltage increases rapidly
   (d) the diode current increases rapidly   (   )

22

2

DIODE RECTIFIER CIRCUITS

PURPOSE AND BACKGROUND

The purpose of this experiment is to demonstrate the characteristics of three different diode rectifier circuits: half-wave rectifier, center-tapped full-wave rectifier, and full-wave bridge rectifier. Each type causes an ac input voltage to be converted into a pulsed waveform having an average, or dc, voltage output.

   Text References: 2–1, Half-Wave Rectifiers; 2–2, Full-Wave Rectifiers.

REQUIRED PARTS AND EQUIPMENT

- ☐ 1-kΩ resistor, 1/2 W
- ☐ Four 1N4001 silicon rectifier diodes
- ☐ 12.6-V rms secondary center-tapped transformer
- ☐ Dual trace oscilloscope
- ☐ VOM or DMM
- ☐ Breadboarding socket

USEFUL FORMULAS

Half-wave rectifier

(1) dc voltage output $= \dfrac{V_S - V_B}{\pi}$ (sine wave input)

(2) Diode PIV $= V_p$

(3) Output frequency $=$ input frequency

Center-tapped full-wave rectifier

(4) dc voltage output $= \dfrac{2V_S - V_B}{\pi}$ (sine wave input)

(5) Diode PIV $-2V_p$

(6) Output frequency $= 2 \times$ input frequency

Full-wave bridge rectifier

(7) dc voltage output $= \dfrac{2(V_S - V_B)}{\pi}$ (sine wave input)

(8) Diode PIV $= V_p$

(9) Output frequency $= 2 \times$ input frequency

PROCEDURE

1. Wire the half-wave rectifier circuit shown in Figure 2–1A, paying careful attention to the polarity of the 1N4001 diode. You should be very careful to be sure that your connections to the 117-V primary of the transformer are properly protected so that you will not get a shock by accidentally touching them. Furthermore, you should have a 1/2-A fuse on the primary side of the transformer. Note that neither of the transformer's primary leads is grounded, while the center-tapped secondary lead is not used in this section!

2. Set your oscilloscope to the following approximate settings:

   Channels 1 and 2: 10 V/division, dc coupling
   Time base: 5 ms/division

   Apply 117 VAC (rms) to the transformer's primary leads. Connect one scope probe to the anode terminal of the 1N4001 diode (point A), and the other probe to the diode's cathode terminal (point B). If everything is working properly, you should obtain the waveforms shown in Figure 2–2.

3. Measure the transformer's peak secondary voltage (V_S), as well as the peak voltage (V_p) across the 1-kΩ resistor, recording your results in Table 2–1. Are the two readings the same?

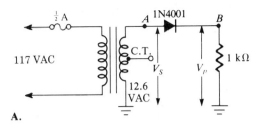

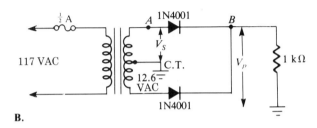

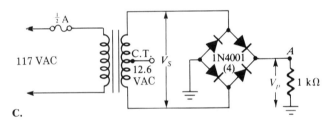

FIGURE 2–1 *Schematic diagram of circuits.*

You should find that these two readings differ slightly. The voltage difference is the *barrier potential* of the diode (V_B), which is approximately 0.7 V for silicon diodes. When the peak voltage is at least ten times larger than this diode voltage drop, the barrier potential usually can be safely neglected, so that these two readings can be considered essentially the same.

4. With your VOM or DMM, measure the *dc* voltage (V_{DC}) across the 1-kΩ resistor, and record your result in Table 2–1. Compare this result with that obtained from the equation for the average or dc voltage of a half-wave rectifier (Equation 1).

   Observe both waveforms. Notice that the frequency of the rectified output sine wave is the same as that of the input sine wave, even though half of each cycle of the output is zero. Why?

5. Turn off the power to the transformer, and wire the center-tapped full-wave rectifier circuit shown in Figure 2–1B. Again, pay careful attention to the polarity of both diodes and the

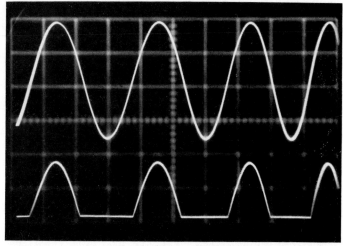

Point *A*
10 V/division

Point *B*
10 V/division

FIGURE 2–2 *Time base: 5 ms/division.*

connections to the 117-V primary of the transformer. The center-tapped lead is grounded for this section.

6. Now set your oscilloscope to the following approximate settings:

Channels 1 and 2: 5 V/division, dc coupling
Time base: 5 ms/division

Apply 117 VAC (rms) to the transformer's primary leads. Connect one probe to the anode terminal of the 1N4001 diode (point *A*), and the other probe to one of the diode's cathode terminals (point *B*). If everything is working properly, you should obtain the waveforms as shown in Figure 2–3.

7. Measure the transformer's peak secondary voltage (V_S) with respect to the grounded center tap, as well as the peak voltage (V_p) across the 1-kΩ resistor, recording your results in Table 2–1. How do these readings compare with those of Step 3?

The peak secondary voltage should be half that of Step 3.

8. With your VOM or DMM, measure the *dc* voltage (V_{DC}) across the 1-kΩ resistor, and record your result in Table 2–1. Compare this result with that obtained from the equation for the average or dc voltage of a center-tapped full-wave rectifier (Equation 4).

Observe both waveforms. Notice that the frequency of the rectified output sine wave is now twice that of the input sine wave. Why?

9. Turn off the power to the transformer, and wire the full-wave bridge rectifier circuit shown in Figure 2–1C. Pay careful attention to the polarity of all four diodes and the connections to the 117-V primary of the transformer. The center-tapped lead is not used for this section. Remove the oscilloscope probe from the anode of the diode.

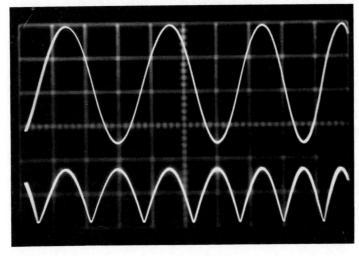

Point *A*
5 V/division

Point *B*
5 V/division

FIGURE 2–3 *Time base: 5 ms/division.*

10. Apply 117 VAC (rms) to the transformer's primary leads. With the channel set to dc coupling, connect only the probe to the ungrounded lead of the 1-kΩ resistor (point *A*). If everything is working properly, you should obtain the same full-wave rectified waveform obtained in Step 6.

11. Measure the peak voltage (V_p) across the 1-kΩ resistor, recording your result in Table 2–1. How does this reading for V_p compare with those of Steps 3 and 7?

    The peak secondary voltage should be the same as that of Step 3 and *twice* that of Step 7. In addition, you should find that the peak voltage across the 1-kΩ resistor is smaller than the secondary voltage by twice the barrier potential. Why?

12. With your VOM or DMM, measure the *dc* voltage (V_{DC}) across the 1-kΩ resistor, and record your result in Table 2–1. Compare this result with that obtained from the equation for the average or dc voltage of a full-wave bridge rectifier (Equation 7).

    Observe both waveforms. Notice that the frequency of the rectified output sine wave is twice that of the input sine wave. Why?

WHAT YOU HAVE DONE

This experiment compared the output characteristics of three types of rectifier circuits: half-wave rectifier, full-wave rectifier using a center-tapped transformer secondary, and a full-wave bridge rectifier. Each converts an ac voltage into a pulsed waveform having an average or dc voltage output.

NOTES

Name ______________________________ Date ____________

DIODE RECTIFIER CIRCUITS

OBJECTIVES/PURPOSE:

SCHEMATIC DIAGRAM:

Name _____   Date _____

DATA FOR EXPERIMENT 2

TABLE 2–1

Measured Parameter	Half-Wave Rectifier	Center-Tapped Full-Wave Rectifier	Bridge Full-Wave Rectifier
V_S			Same as center-tapped full-wave rectifier
V_p			
V_{DC}			

Name _____ Date _____

RESULTS AND CONCLUSIONS:

REVIEW QUESTIONS FOR EXPERIMENT 2

1. For the half-wave rectifier circuit of Figure 2–1A, the peak load
   voltage is approximately
   (a) 6 V   **(b)** 12 V   **(c)** 18 V   **(d)** 24 V   (   )
2. For an input frequency of 60 Hz, the period of the half-wave
   signal is approximately
   (a) 4 ms   **(b)** 8 ms   **(c)** 16 ms   **(d)** 32 ms   (   )
3. Compared to the dc output voltage of the half-wave rectifier
   of Figure 2–1A, the dc output voltage of the full-wave bridge
   rectifier of Figure 2–1C is approximately
   (a) one-half as large   **(b)** the same   **(c)** twice as large   (   )
4. In this experiment, the rectifier circuit that has the lowest diode
   peak inverse voltage is the
   (a) half-wave rectifier   **(b)** full-wave center-tapped rectifier
   (c) full-wave bridge rectifier   **(d)** both a and c   (   )
5. In this experiment, the rectifier circuit that has the greatest dc
   output voltage is the
   (a) half-wave rectifier   **(b)** full-wave center-tapped rectifier
   (c) full-wave bridge rectifier   (   )

NOTES

3

THE CAPACITOR INPUT
RECTIFIER FILTER

PURPOSE AND BACKGROUND

The purpose of this experiment is to demonstrate the operation of a capacitor input filter when connected to the output of a full-wave bridge rectifier. The filter, which consists of a single resistor and capacitor in parallel, smooths out the pulsating output voltage of the rectifier.

Text References: 2–3, Power Supply Filters; Appendix B, Derivations.

REQUIRED PARTS AND EQUIPMENT

☐ 1-kΩ resistor, 1/2 W
Capacitors (25 V):
    ☐ 100 μF
    ☐ 470 μF
☐ Four 1N4001 silicon
  rectifier diodes

☐ 12.6-V rms secondary
  center-tapped transformer
☐ Dual trace oscilloscope
☐ VOM or DMM
☐ Breadboarding socket

USEFUL FORMULAS

dc output voltage

$$(1) \quad V_{dc} = \left(1 - \frac{0.00417}{R_L C}\right) V_{p(in)} \quad \text{(when } R_L C \gg T_{input})$$

rms ripple voltage

$$(2) \quad V_r = \frac{0.0024}{R_L C} V_{p(in)} \quad \text{(when } R_L C \gg T_{input})$$

Percent ripple factor

$$(3) \quad \%r = \frac{V_r}{V_{dc}} \times 100\%$$

PROCEDURE

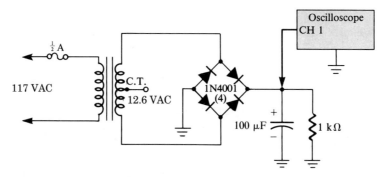

FIGURE 3–1 *Schematic diagram of circuit.*

1. Wire the full-wave bridge rectifier circuit shown in Figure 3–1, paying careful attention to the polarity of the 1N4001 diodes. You should be very careful to be sure that your connections to the 117-V primary of the transformer are properly protected so that you will not get a shock by accidentally touching them. Furthermore, you should have a 1/2-A fuse on the primary side of the transformer. Note that neither of the transformer's primary leads is grounded, while the center-tapped secondary lead is not used. Also observe the polarity of the 100-μF filter capacitor.

2. Apply 117 VAC (rms) to the transformer's primary leads. With one oscilloscope channel set to dc coupling, connect the probe to the ungrounded junction of the 1-kΩ resistor and the 100-μF capacitor. If everything is working properly, you should obtain the waveform shown in Figure 3–2.

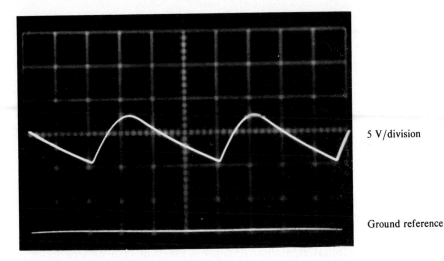

5 V/division

Ground reference

FIGURE 3–2 *Time base: 2 ms/division.*

3. With the oscilloscope, measure the peak output voltage (V_p) across the 1-kΩ resistor and the 100-μF capacitor. With your VOM or DMM, measure the dc voltage (V_{dc}) and compute the expected dc voltage, rms ripple, and percentage ripple factor using Equations 1, 2, and 3. Record all results in Table 3–1.

4. Turn off the 177-VAC primary voltage, and then place a piece of wire or a screwdriver across both capacitor leads, which, in effect, discharges the capacitor. With the relatively low secondary voltages used in this experiment, the risk of getting a severe shock is small. However, this practice is a good habit to acquire when working with power supplies and filters. By discharging (that is, shorting) the filter capacitor *after* the supply voltage has been turned off or removed, you then eliminate the possibility of coming in contact with a fully charged capacitor, which, depending on its capacitance and voltage, can deliver quite an unexpected jolt.

   Remove the 100-μF capacitor from the circuit and replace it with a 470-μF capacitor. Then apply 117 VAC to the transformer's primary.

5. With your oscilloscope, measure the peak output voltage across the 1-kΩ resistor and the 470-μF capacitor. As in Step 3, measure the dc output voltage and, using Equations 1, 2, and 3, calculate the expected values for the dc voltage, rms ripple voltage, and percent ripple factor. Record all results in Table 3–1.

6. If you have wired the circuit correctly, you should now observe very little ripple voltage on the oscilloscope's display. Now change the input to *ac* coupling, and increase the sensitivity of

the oscilloscope channel to about 10 mV/division so that you can clearly see the output ripple.

7. For each capacitor value, compare your values for dc output voltage, rms ripple voltage, and percent ripple factor. When you increase the value of the filter capacitor, what happens to the dc output voltage, rms ripple voltage, and percent ripple factor?

   For a fixed load resistance of 1 kΩ, increasing the capacitance of the input filter capacitor should increase the dc output voltage toward the peak output voltage while decreasing both the rms ripple voltage and the percent ripple factor.

WHAT YOU HAVE DONE

This experiment demonstrated the operation of a capacitor input filter when connected to the output of a full-wave bridge rectifier. The filter, using a parallel resistor-capacitor circuit, smooths out the pulsating output voltage of the rectifier. As the RC time constant of the filter was made larger, the ripple voltage of the filter was reduced further.

Name _____ Date _____

THE CAPACITOR INPUT RECTIFIER FILTER

OBJECTIVES/PURPOSE:

SCHEMATIC DIAGRAM:

Name _____ Date _____

DATA FOR EXPERIMENT 3

TABLE 3–1

Parameter	Step 3	Step 5
Secondary peak voltage, V_p: Measured		
dc output voltage, V_{dc}: Calculated		
Measured		
rms ripple voltage, V_r: Calculated		
Measured (peak-to-peak)		
% ripple factor, % r: Calculated		

Name _____ Date _____

RESULTS AND CONCLUSIONS:

REVIEW QUESTIONS FOR EXPERIMENT 3

1. For the circuit of Figure 3–1, the time constant of the capacitor input filter is
   (a) 1 ms   **(b)** 10 ms   **(c)** 100 ms   **(d)** 1 s   (   )
2. In Step 5, the dc output voltage is approximately
   (a) 6 V   **(b)** 12 V   **(c)** 18 V   **(d)** 24 V   (   )
3. As the time constant of the input filter is decreased, the dc output voltage
   (a) decreases   **(b)** increases   **(c)** remains the same   (   )
4. As the time constant of the input filter is decreased, the output ripple voltage
   (a) decreases   **(b)** increases   **(c)** remains the same   (   )
5. In a well-designed power supply, the percent ripple factor should be
   (a) close to 0%   **(b)** approximately 50%
   (c) close to 100%   (   )

39

NOTES

4

THE DIODE LIMITER

PURPOSE AND BACKGROUND

The purpose of this experiment is to demonstrate the operation of a diode limiter. Diode limiters are wave-shaping circuits in that they are used to prevent signal voltages from going above or below certain levels. The limiting level may be either equal to the diode's barrier potential or made variable with a dc source voltage. Because of this clipping capability, the limiter is also called a *clipper*.

    Text Reference: 2–4, Diode Limiting and Clamping Circuits.

REQUIRED PARTS AND EQUIPMENT

- [ ] 15-kΩ resistor, 1/4 W
- [ ] 5-kΩ potentiometer, or 10-turn "trimpot"
- [ ] 1N4001 silicon rectifier diode

- [ ] 0–15 V dc power supply
- [ ] Signal generator
- [ ] Dual trace oscilloscope
- [ ] Breadboarding socket

PROCEDURE

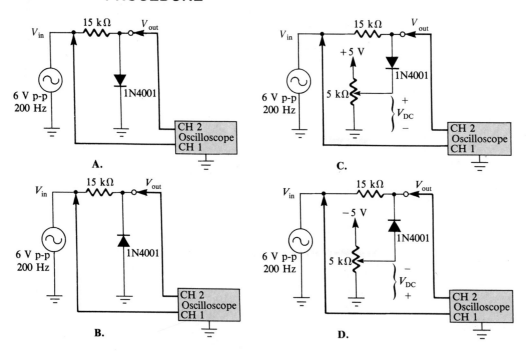

FIGURE 4–1 *Schematic diagram of circuits.*

1. Wire the limiter circuit shown in the schematic diagram of Figure 4–1A. Set your oscilloscope to the following approximate setting:

    > Channels 1 and 2: 1 V/division, dc coupling
    > Time base: 1 ms/division

    Without any input signal connected to the breadboard, position the two lines on the oscilloscope's display so that they are at the same level (that is, zero volts).

2. Now connect the signal generator to the breadboard. Adjust the signal generator's output level at 6 V peak-to-peak at a frequency of 200 Hz. You should see two waveforms similar to those shown in Figure 4–2. Notice that the positive peaks of the limiter's output waveform are removed, or *clipped* off. Notice also that the clipping level is not perfect; the positive peaks are clipped not at zero volts, but at a small positive voltage. When the input waveform goes positive at a level greater than the barrier potential of the diode, the diode is forward biased, the equivalent of a short circuit in series with a small dc voltage source. Thus, approximately 0.5 to 0.7 volt (the barrier potential

for a silicon diode) is dropped across the diode. When the input waveform goes negative, the diode looks like an open circuit, and essentially all of the input appears at the output. Such an arrangement is called a *positive* limiter because the circuit limits the positive peaks of the input waveform. On the data page at the end of this experiment, sketch your clipped waveform, showing the positive and negative peak values.

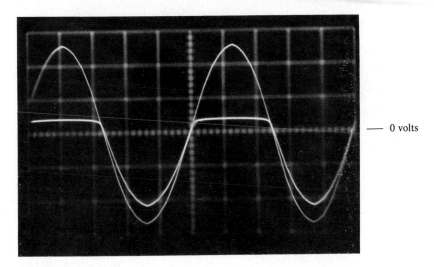

— 0 volts

FIGURE 4–2

3. Now reverse the polarity of the diode in the circuit, as shown in Figure 4–1B. How does this waveform compare with that of Step 2?

  The behavior is opposite that of the positive limiter. The waveform has all *negative* peaks of the input signal removed, as shown in Figure 4–3. Again, notice that the clipping level is not perfect; the negative peaks are clipped not at zero volts, but at a small negative voltage. Such an arrangement is called a *negative* limiter because the circuit clips off the negative peaks of the input waveform. On the data page at the end of this experiment, sketch your clipped waveform, showing the positive and negative peak values.

4. Now connect the circuit of Figure 4–1C. Apply power to the breadboard and adjust the potentiometer so that the dc voltage (V_{DC}) is +1.5 V. Connect the signal generator, set at 6 V peak-to-peak, to the breadboard. What do you notice about the output of the limiter?

  The clipping level is *higher* than that measured in Step 2. The circuit uses a dc source voltage to *bias,* or set, the clipping

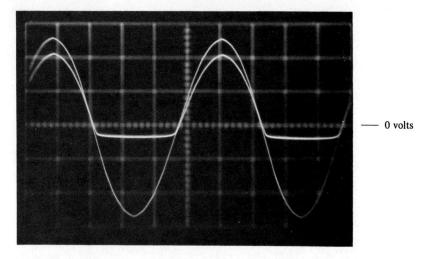

—— 0 volts

FIGURE 4–3

level. Consequently this arrangement is called a *positive-biased* limiter. On the data page at the end of this experiment, sketch the clipped waveform, showing the dc positive and negative peak values.

Note that the positive clipping level is the dc source voltage *plus* the diode's barrier potential. For the diode to become forward biased, the positive peaks of the input signal must be greater than the dc source voltage and the diode's barrier potential.

5. Vary the resistance of the potentiometer from one extreme to the other. What happens to the clipping level?

The clipping level changes with the setting of the potentiometer. At one extreme, when the dc bias voltage (V_{DC}) is zero, the positive clipping level should be the same as was measured in Step 2. At the other extreme, there should be no clipping, as the dc bias voltage is about +5 V. Since the input positive peaks are at +3.0 V, the diode is effectively reverse biased and looks like an open circuit, and thus the input appears unchanged at the output.

6. Now reverse the polarities of both the diode and the dc power supply in the circuit, as shown in Figure 4–1D. Adjust the potentiometer so that the dc voltage (D_{DC}) is −1.5 V. Connect the signal generator, set at 6 V peak-to-peak, to the breadboard. What do you notice about the output of the limiter?

Note that the clipping level is *lower* than that measured in Step 3. The circuit uses a dc source voltage to *bias,* or set, the clipping level. Consequently, this arrangement is called a *negative-biased* limiter. On the data page at the end of this experiment, sketch the clipped waveform, showing the dc positive and negative peak values.

Notice also that the negative clipping level is the dc source voltage plus the diode's barrier potential. For the diode to become forward biased, the negative peaks of the input signal must be greater than the dc source voltage and the diode's barrier potential.

7. Vary the resistance of the potentiometer from one extreme to the other. What happens to the clipping level?

The clipping level changes with the setting of the potentiometer. At one extreme, when the dc bias voltage (V_{DC}) is zero, the positive clipping level should be the same as was measured in Step 3. At the other extreme, there should be no clipping, as the dc bias voltage is about -5 V. Since the input negative peaks are at -3.0 V, the diode is effectively reverse biased and looks like an open circuit, and thus the input appears unchanged at the output.

WHAT YOU HAVE DONE

This experiment demonstrated the operation of a diode limiter, or *clipper*, which limits signal voltages from going above or below preset levels. You worked with both positive and negative limiters whose clipping level was equal to the diode's barrier potential. In addition, it was shown how to make the clipping level variable by using an external dc voltage source.

NOTES

Name ______________________ Date __________

THE DIODE LIMITER

OBJECTIVES/PURPOSE:

SCHEMATIC DIAGRAM:

NOTES

Name _____ Date _____

DATA FOR EXPERIMENT 4

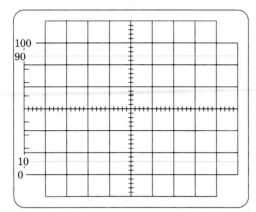

Volts/div= _____ Time/div=_____
Positive clipper (Step 2)

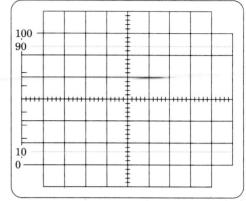

Volts/div=_____ Time/div=_____
Negative clipper (Step 3)

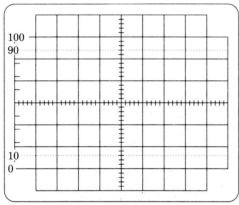

Volts/div=_____ Time/div=_____
Positive-biased clipper (Step 4)

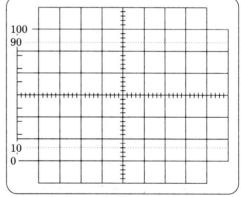

Volts/div=_____ Time/div=_____
Negative biased clipper (Step 6)

Name _____ Date _____

RESULTS AND CONCLUSIONS:

REVIEW QUESTIONS FOR EXPERIMENT 4

1. For the positive limiter circuit of Figure 4–1A, the positive peak
    voltage is approximately
    (a) 0 V   **(b)** +0.6 V   **(c)** +3 V   **(d)** +6 V   (   )
2. For the negative limiter circuit of Figure 4–1B, the positive
    peaks are not clipped because the diode is
    (a) reverse biased   **(b)** forward biased   (   )
3. In all the limiting circuits in this experiment, the 15-kΩ resistor
    is used to
    (a) set the clipping level
    (b) set the peak output voltage
    (c) limit the voltage across the diode
    (d) limit the peak forward diode current   (   )
4. For the circuit of Figure 4–1C, the potentiometer is used to set
    the clipping level of the output's
    (a) positive peaks   **(b)** negative peaks
    (c) positive and negative peaks   (   )
5. For the circuit of Figure 4–1D, the potentiometer is used to set
    the clipping level of the output's
    (a) positive peaks   **(b)** negative peaks
    (c) positive and negative peaks   (   )

5

THE DIODE CLAMPER

PURPOSE AND BACKGROUND

The purpose of this experiment is to demonstrate the operation of a diode clamper. Like the diode clipper, the clamper is a wave-shaping circuit, but it adds a dc level to the input waveform. Thus, the clamper is often referred to as a *dc restorer*. However, unlike that of the clipper, the *shape* of the input signal of a clamper is not changed.

   Text Reference: 2–4, Diode Limiting and Clamping Circuits.

REQUIRED PARTS AND EQUIPMENT

- ☐ 10-kΩ resistor, 1/4 W
- ☐ 10-μF electrolytic capacitor, 25 V
- ☐ 1N4001 silicon rectifier diode
- ☐ Signal generator
- ☐ Dual trace oscilloscope
- ☐ Breadboarding socket

USEFUL FORMULAS

Clamper time constant

(1) $10R_L C \gg T_{\text{input}}$

Peak output voltage

(2) $V_{\text{out}}(\text{peak}) = V_{\text{in}} (\text{peak-to-peak}) - V_d$

PROCEDURE

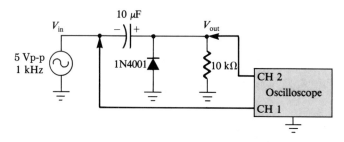

FIGURE 5–1 *Schematic diagram of circuit.*

1. Wire the clamper circuit shown in the schematic diagram of Figure 5–1. Set your oscilloscope to the following approximate settings:

   Channels 1 and 2: 2.0 V/division, dc coupling
   Time base: 0.2 ms/division

   Without any input signal connected to the breadboard, position the two lines on the oscilloscope's display so that they are at the same level (that is, zero volts).

2. Now connect the signal generator to the breadboard. Adjust the signal generator's output level at 5 V peak-to-peak at a frequency of 1 kHz. You should see two sine waves. Notice that the clamper's output signal level is above the input's. This action is that of a *positive* clamper, so the input waveform is shifted upward. This effect is the same as that obtained by adding a dc voltage onto the input waveform. On the data page at the end of this experiment, sketch both the input and the output waveforms, showing the positive and negative peak values for both.

3. Note that the clamping action is not perfect. The negative peaks of the output waveform are clamped not at zero volts, but at a small negative voltage. When the input waveform goes negative at a level greater than the barrier potential of the diode, the diode is forward biased, the equivalent of a short circuit

52

in series with a small dc voltage source. Thus, approximately 0.5 to 0.7 volt (the barrier potential for a silicon diode, V_d) is dropped across the diode, while the remainder of the peak negative voltage ($V_p - V_d$) charges the 10-μF capacitor. On the next positive-going half-cycle, the diode is reverse biased, looking like an open circuit, and the voltage stored on the capacitor is then added to the time-varying input voltage. The result is that the peak output voltage is now approximately equal to the peak-to-peak input voltage, less the voltage drop of the diode.

4. Increase the peak-to-peak input voltage. What happens?

Although the peak-to-peak output voltage increases, its negative peak remains clamped at the same negative voltage level measured in Step 3. You should find that the positive peak output voltage is again approximately equal to the peak-to-peak input voltage.

5. Now reverse the polarity of the diode in the circuit, and repeat Steps 2, 3, and 4. Now what happens?

The behavior is opposite that of the positive clamper. Notice that the clamper's output signal level is *below* the input's. This action is that of a *negative* clamper, so the input waveform is shifted downward. This effect is the same as that obtained by adding a negative dc voltage onto the input waveform. On the data page at the end of this experiment, sketch both the input and the output waveforms, showing the positive and negative peak values for both.

6. Again you should notice that the clamper action is not perfect. The positive peaks of the output waveform are clamped not at zero volts, but at a small positive voltage.

7. Increase the peak-to-peak input voltage. What happens?

You should see that although the peak-to-peak output voltage increases, its positive peak remains clamped at the same positive voltage level measured in Step 6. You should find that the negative peak output voltage is again approximately equal to the peak-to-peak input voltage.

WHAT YOU HAVE DONE

This experiment demonstrated the operation of a diode clamper. This circuit does not change the waveshape of the input signal, but merely adds a dc level to the input waveform.

NOTES

Name _____   Date _____

THE DIODE CLAMPER

OBJECTIVES/PURPOSE:

SCHEMATIC DIAGRAM:

NOTES

Name _____ Date _____

DATA FOR EXPERIMENT 5

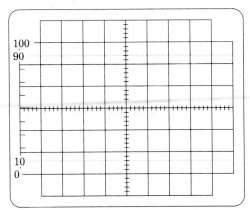

Volts/div=_____ Time/div=_____
Positive clamper (step 2)

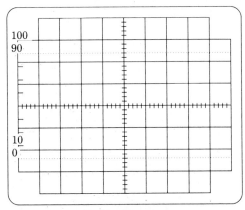

Volts/div=_____ Time/div=_____
Negative clamper (step 5)

Name _____ Date _____

RESULTS AND CONCLUSIONS:

REVIEW QUESTIONS FOR EXPERIMENT 5

1. For the circuit of Figure 5–1 to function properly, the input frequency should be at least
   (a) 1 Hz   **(b)** 10 Hz   **(c)** 100 Hz   **(d)** 1 kHz   (   )
2. For the circuit of Figure 5–1, if the input signal has a peak voltage of V_p, then the output signal is
   (a) shifted upward by approximately V_p
   (b) shifted upward by approximately $2V_p$
   (c) shifted downward by approximately V_p
   (d) shifted downward by approximately $2V_p$ (   )
3. For the circuit of Figure 5–1, the negative peak voltage of the output signal is approximately
   (a) $-V_p$
   (b) -0.7 V
   (c) 0 V
   (d) $+0.7$ V   (   )
4. If the peak-to-peak input voltage is increased,
   (a) the peak-to-peak output voltage remains approximately equal to the peak-to-peak input voltage
   (b) the negative peak output voltage remains clamped at approximately -0.7 V
   (c) the output peak voltage approximately equals the peak-to-peak input voltage
   (d) all of the above   (   )
5. In order to change the circuit of Figure 5–1 to a negative clamper, you must
   (a) reverse the polarity of the signal source
   (b) reverse the polarity of the diode
   (c) reverse the polarity of the capacitor
   (d) all of the above   (   )

58

6

THE DIODE
VOLTAGE DOUBLER

PURPOSE AND BACKGROUND

The purpose of this experiment is to demonstrate the operation of both the half-wave and the full-wave diode voltage doublers. Diode voltage doublers are used to double the peak rectified voltage without the necessity of increasing the input transformer's voltage rating. The half-wave voltage doubler is actually a positive clamper followed by a half-wave rectifier with a capacitor input filter (peak detector). It charges a series capacitor on each positive half-cycle. Consequently, the ripple frequency is the same as the input frequency.

The full-wave voltage doubler, on the other hand, has the same rectified peak output voltage as the half-wave doubler. It charges one of two series capacitors on the first half-cycle, while the other capacitor is charged on the remaining half-cycle. Therefore, it has a ripple frequency that is twice the input frequency. Consequently, for the same filter time constant, the peak-to-peak ripple voltage is smaller when a full-wave voltage doubler is used.

Text Reference: 2–5, Voltage Multipliers.

REQUIRED PARTS AND EQUIPMENT

- ☐ 10-kΩ resistor, 1/4 W
- ☐ Two 100-μF capacitors, 25 V
- ☐ Two 1N4001 silicon rectifier diodes

- ☐ 12.6-V rms secondary center-tapped transformer
- ☐ Dual trace oscilloscope
- ☐ VOM or DMM
- ☐ Breadboarding socket

USEFUL FORMULAS

Output voltage

$$(1)\quad V_o = 2V_s - 2V_d$$

Ripple frequency

$$(2)\quad f_{\text{ripple}} = f_{\text{in}} \qquad \text{(half-wave doubler)}$$

$$(3)\quad f_{\text{ripple}} = 2f_{\text{in}} \qquad \text{(full-wave doubler)}$$

Diode peak inverse voltage

$$(4)\quad \text{PIV} = 2V_s$$

PROCEDURE

1. Wire the half-wave diode voltage doubler shown in the schematic diagram of Figure 6–1A.
2. Set your oscilloscope to the following approximate settings:

   Channels 1 and 2: 5 V/division, dc coupling
   Time base: 2 ms/division

   Apply the 117-V rms ac, 60-Hz power line voltage to the transformer's primary.
3. You should see two sine waves on the oscilloscope's display. On Channel 1 at point *A*, it should show the secondary voltage of the transformer. Measure the positive peak voltage V_s and the frequency f_{in}, recording these values in Table 6–1.

   On Channel 2, you should see the same waveform at point *B,* but it should be positively clamped near zero volts and the positive peak voltage should be nearly twice that of the transformer's secondary voltage. Measure this peak voltage, V_p, and record this value in Table 6–1.
4. Now take the Channel 1 probe and connect it to the 10-kΩ load resistor (point *C*). You should see two signals similar to those shown in Figure 6–2. Measure the dc voltage V_{DC} across the 10-kΩ resistor with a VOM or DMM, and record this value in

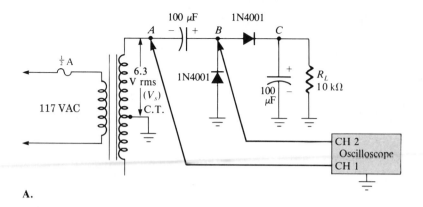

A.

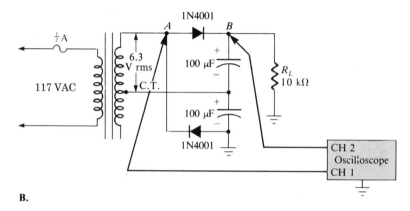

B.

FIGURE 6–1 *Schematic diagram of circuits.*

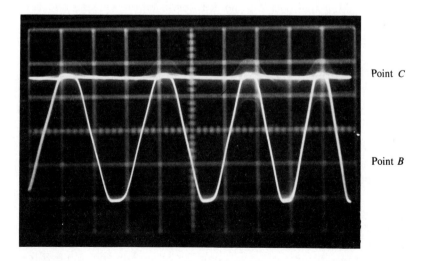

Point *C*

Point *B*

FIGURE 6–2

Table 6–1. You should have measured a dc voltage that is nearly twice that of the transformer's peak secondary voltage, less two diode voltage drops.

5. Now switch Channel 1 to ac coupling, and increase the sensitivity to 0.05 V/division to display adequately the output ripple voltage. Measure both the peak-to-peak ripple voltage and the ripple frequency. Record both values in Table 6–1. You should find that both the ripple frequency and the input power line frequency are the same.

6. Disconnect the power line voltage from the transformer, and wire the full-wave diode voltage doubler shown in the schematic diagram of Figure 6–1B.

7. Set your oscilloscope to the following approximate settings:

> Channels 1 and 2: 5 V/division, dc coupling
> Time base: 2 ms/division

Apply the 117-V rms ac, 60-Hz power line voltage to the transformer's primary.

8. You should see two sine waves on the oscilloscope's display. On Channel 1 at point A, you should see the transformer's secondary voltage positively clamped near zero volts. Measure its positive peak voltage V_p and its frequency f_{in}, recording these values in Table 6–2. Measure the dc voltage V_{dc} across the 10-kΩ resistor with a VOM or DMM, and record this value in Table 6–2. You should have measured a dc voltage that is nearly twice that of the transformer's peak secondary voltage, less two diode voltage drops.

9. Now switch Channel 2 to ac coupling and increase the sensitivity to 0.05 V/division to display adequately the output ripple voltage. Measure both the peak-to-peak ripple voltage and the ripple frequency. Record both values in Table 6–2. You should find that the ripple frequency is twice that of the power line input and that the peak-to-peak ripple voltage is smaller than that of the half-wave doubler circuit using the same capacitor and resistor values. These differences occur because the capacitors are charged and partially discharged twice as fast as they are in a half-wave doubler.

WHAT YOU HAVE DONE

This experiment demonstrated the operation of half-wave and full-wave diode voltage doublers. The half-wave doubler was actually a positive clamper circuit followed by a half-wave rectifier with a capacitor input filter acting as peak detector. For the same filter RC time constant, the peak-to-peak ripple voltage of the full-wave doubler is smaller than the half-wave circuit.

Name ______________________________ Date ____________

THE DIODE VOLTAGE DOUBLER

OBJECTIVES/PURPOSE:

SCHEMATIC DIAGRAM:

DATA FOR EXPERIMENT 6

TABLE 6–1 *Half-wave diode voltage doubler.*

Parameter	Measured Value
V_s	V
f_{in}	Hz
V_p	V
V_{dc}	V
f_{ripple}	Hz
V_{ripple}	V

TABLE 6–2 *Full-wave diode voltage doubler.*

Parameter	Measured Value
V_p	V
f_{in}	Hz
V_{dc}	V
f_{ripple}	Hz
V_{ripple}	V

Name _____ Date _____

RESULTS AND CONCLUSIONS:

REVIEW QUESTIONS FOR EXPERIMENT 6

1. For the half-wave voltage doubler of Figure 6–1A, the output
    ripple frequency is
    (a) one-half the input frequency
    (b) the same as the input frequency
    (c) twice the input frequency   (   )
2. For the full-wave voltage doubler of Figure 6–1B, the output
    ripple frequency is
    (a) one-half the input frequency
    (b) the same as the input frequency
    (c) twice the input frequency   (   )
3. If the peak input voltage is 10 V, then the peak inverse voltage
    of both diodes of a half-wave voltage doubler is
    (a) 5 V   (b) 10 V   (c) 20 V   (d) 40 V   (   )
4. If the peak input voltage is 10 V, then the peak inverse voltage
    of both diodes of a full-wave voltage doubler is
    (a) 5 V   (b) 10 V   (c) 20 V   (d) 40 V   (   )
5. For the half-wave voltage doubler of Figure 6–1A, the capacitor
    and diode arrangement between points A and B is a
    (a) peak detector   (b) negative limiter
    (c) positive clamper   (d) negative clamper   (   )

NOTES

7

THE ZENER DIODE AND VOLTAGE REGULATION

PURPOSE AND BACKGROUND

The purposes of this experiment are to demonstrate (1) the characteristics of a zener diode and (2) its use as a simple voltage regulator. Unlike rectifier diodes, zener diodes are normally *reverse biased,* so they maintain a constant voltage across their terminals over a specified range of current. Like a rectifier diode, a zener diode can be approximated by a constant dc voltage source in series with a resistor. When used as a regulator, the zener diode maintains a dc output voltage that is essentially constant even though the load current may vary.

Text References: 3–1, Zener Diodes; 3–2, Zener Diode Applications.

REQUIRED PARTS AND EQUIPMENT

Resistors:
- [ ] 100 Ω, 1/4 W
- [ ] Two 220 Ω, 1/2 W
- [ ] 1N753, 6.2-V, 400-mW zener diode
- [ ] 0–15 dc power supply

- [ ] Signal generator
- [ ] Two DMMs (preferred) or VOMs
- [ ] Dual trace oscilloscope
- [ ] Breadboarding socket

67

USEFUL FORMULAS

Maximum limiting series resistance

$$(1) \ R_s(\text{max}) = \frac{V_{\text{in}}(\text{min}) - V_{\text{out}}}{I_L(\text{max})}$$

Output voltage

$$(2) \ V_{\text{out}} = V_Z \qquad (\text{ideal})$$

$$(3) \ V_{\text{out}} = V_Z + I_Z R_Z \qquad (\text{actual})$$

where I_Z = zener diode current

R_Z = zener diode internal resistance = $\Delta V_Z / \Delta I_Z$

Zener diode current

$$(4) \ I_Z = I_S - I_L$$

Source current

$$(5) \ I_S = \frac{V_{\text{in}} - V_{\text{out}}}{R_S}$$

Zener diode power dissipation

$$(6) \ P_Z = I_Z V_Z$$

Percent load regulation

$$(7) \ \%\text{VR} = \frac{V_{\text{NL}} - V_{\text{FL}}}{V_{\text{FL}}} \times 100\%$$

where V_{NL} = no-load (open circuit) output voltage

V_{FL} = full-load output voltage

Output ripple voltage

$$(8) \ V_o(\text{ripple}) = \left[\frac{R_L \| R_Z}{R_S + (R_L \| R_Z)} \right] V_{\text{in}}(\text{ripple})$$

PROCEDURE

1. Wire the circuit shown in the schematic diagram of Figure 7–1A (p. 60).
2. Increase the dc supply voltage in small steps while simultaneously measuring the voltage across (V_Z) and the current through (I_Z) the zener diode. In the vicinity of the zener's knee voltage (approximately 6 V), make these steps approximately 0.05 V. *Do not exceed a zener current of 40 mA.* Record your data in Table 7–1A, and plot your results for the corresponding zener current and voltage values on the graph provided

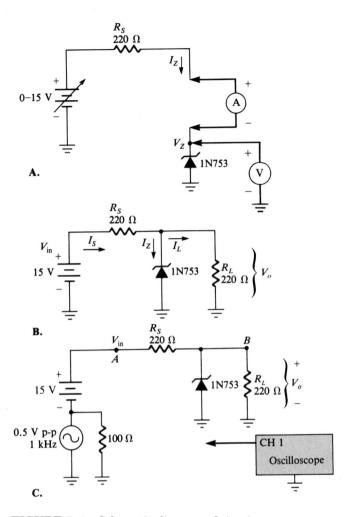

A.

B.

C.

FIGURE 7–1 *Schematic diagram of circuits.*

for this purpose. What do you notice about the current-voltage curve for the zener diode?

Note that initially, the zener diode current is essentially zero for diode voltages less than the knee voltage. You should find that as the voltage drop approaches the diode's knee voltage, the diode's current increases rapidly, while, at the same time, the voltage stays essentially constant. Consequently, the zener diode maintains an essentially constant voltage drop when it is sufficiently reverse biased.

The 1N753 diode is rated at 6.2 V with a tolerance of 10%. From your graph, determine the voltage across the zener diode at a current of approximately 20 mA. Within 10%, your value should be 6.2 V. Record the measured zener voltage in Table 7–1B.

3. Determine the internal resistance R_Z of your 1N753 zener diode from your data by taking the change in zener voltage, ΔV_Z, divided by the corresponding change in current, ΔI_Z. Do this calculation only on the straight-line breakdown region of your diode curve that you plotted in Step 2. Record your result for the internal zener resistance in Table 7–1B.

4. Now wire the circuit shown in the schematic diagram of Figure 7–1B.

5. Apply dc voltage (V_{in}) to the breadboard. Measure the source current (I_S), zener current (I_Z), load current (I_L), and full-load output voltage V_{FL}, recording your values in Table 7–2A. Using the zener voltage and the internal zener resistance calculated in Steps 2 and 3, compare the measured output voltage with the expected value (Equation 3).

6. Now disconnect the 220-Ω load resistor. Measure the source current (I_S), zener current (I_Z), and output voltage with no load V_{NL}, recording your values in Table 7–2B. Using the zener voltage and the internal zener resistance determined in Steps 2 and 3, compare the measured no-load output voltage with the expected value.

7. For this circuit, determine the percent load regulation, and record your result in Table 7–2B.

8. Now add a signal generator in series with the dc voltage source as shown in Figure 7–1C. Adjust the output of the signal generator at 0.5 V peak-to-peak with a frequency of 1 kHz.

9. With your oscilloscope at point A, observe both the dc and the ac voltage levels, using your oscilloscope set on *dc* coupling. You should see a 0.5 V peak-to-peak sine wave superimposed on a 15-V dc level above ground.

10. With your oscilloscope at point B, measure the dc output voltage of the zener diode regulator, recording your value in Table 7–2B. At this point you should see virtually no ripple voltage on the regulator's output signal. How does this voltage compare with that measured in Step 5?

11. Now set your oscilloscope to ac coupling and increase its sensitivity to 5 mV/division. You should now observe a sine wave ripple signal, but now much smaller than the 500-mV input ripple voltage. Measure the output peak-to-peak ripple voltage and compare it with the expected value (Equation 8), recording your results in Table 7–2B.

Notice that the zener diode regulator provides a relatively constant output voltage as long as the input voltage is greater than the zener's knee voltage. If there is any voltage variation or ripple on the input voltage signal, the output remains essentially constant.

WHAT YOU HAVE DONE

This experiment demonstrated the characteristics of a 6.2-V zener diode. The zener diode is normally reverse biased so that it maintains a constant voltage between its anode and cathode terminals over a specified range of current. This experiment demonstrated the concept of voltage regulation where the output voltage remained essentially constant with changes in load current.

NOTES

Name _____ Date _____

THE ZENER DIODE AND VOLTAGE REGULATION

OBJECTIVES/PURPOSE:

SCHEMATIC DIAGRAM:

DATA FOR EXPERIMENT 7

TABLE 7–1 *Zener diode characteristic curve.*

A.

Zener Voltage, V_Z (V)	Zener Current, I_Z (mA)

B.

Zener knee voltage @ $I_Z = 20$ mA	V
Internal zener resistance	Ω

Name _____ Date _____

TABLE 7–2 *Zener diode voltage regulator.*

A. Full-load data

Parameter	Measured Value	Expected Value	% Error
I_S			
I_Z			
I_L			
V_{FL}			

B. No-load data

Parameter	Measured Value	Expected Value	% Error
I_S			
I_Z			
V_{NL}			
% load regulation, % VR			
dc input voltage, V_{IN} (dc)			
ac input ripple voltage, V_{in} (ripple), peak-to-peak			
dc output voltage, V_o (dc)			
ac output ripple voltage, V_o (ripple), peak-to-peak			

NOTES

Name _____ Date _____

DATA FOR EXPERIMENT 7

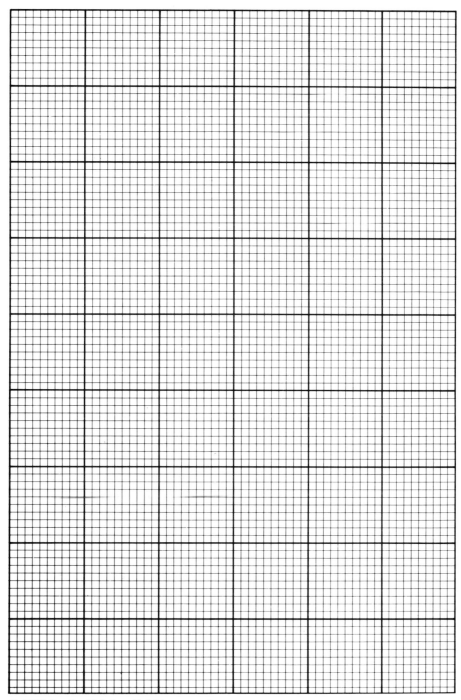

Name _____ Date _____

RESULTS AND CONCLUSIONS:

REVIEW QUESTIONS FOR EXPERIMENT 7

1. From your data, the zener voltage for the zener diode used in this experiment is approximately
   (a) 0.3 V (b) 0.7 V (c) 6 V (d) 10 V   (   )
2. For which portion of the diode curve does the zener diode look like an open circuit?
   (a) Diode voltages less than the zener voltage.
   (b) Diode voltages greater than the zener voltage.   (   )
3. For the circuit of Figure 7–1B, if the input voltage is less than 6 V, the output voltage is
   (a) 0 V (b) 6 V (c) the same as the input   (   )
4. If the load resistor of Figure 7–1B is disconnected, the current through the zener diode is approximately
   (a) 0 mA (b) 10 mA (c) 20 mA (d) 40 mA   (   )
5. The power dissipated by the zener diode for the circuit of Figure 7–1B is greatest when
   (a) the zener diode is shorted
   (b) the load resistor is removed
   (c) the load resistor is shorted
   (d) the input voltage is increased   (   )

8

USING AN OHMMETER TO TEST TRANSISTOR DIODE JUNCTIONS

PURPOSE AND BACKGROUND

The purpose of this experiment is to demonstrate how to test npn and pnp transistors using an ohmmeter. Since a transistor can be represented internally by two diode junctions, an ohmmeter can be used to check each diode junction, as was done in Experiment 1. Thus, there is a simple test for open or shorted diode junctions. If the three terminals are known, then it is possible to determine if a giveng transistor is npn or pnp.

Text Reference: 4–7, Troubleshooting.

REQUIRED PARTS AND EQUIPMENT

- ☐ 1N914 (or 1N4148) diode
- ☐ 2N3904 npn transistor
- ☐ 2N3906 pnp transistor
- ☐ VOM
- ☐ Breadboarding socket (optional)

PROCEDURE

1. Often one can use a VOM to check quickly whether a diode is good or bad. Unless they have a specific function for this purpose, *most DMMs are usually not able to perform this test properly*. When VOMs are used to measure resistances, the polarities of their leads are sometimes reversed from the normal sense. That is, the positive lead is actually wired to the *negative* terminal of the internal battery. In this case, the forward and reverse resistance readings will be the opposite of those indicated in these two steps. When this type of VOM functions as a voltmeter or an ammeter, its leads are internally connected in the normal sense.

   In order to determine the polarity of your ohmmeter's leads, perform a simple test first on a 1N914 diode. Set the ohmmeter's resistance range switch to a low range, such as the "R × 100" range, and then place one meter lead on the diode's anode lead and the other meter lead on the diode's cathode lead. Note whether the resistance reading is high or low. Then reverse the leads and note if the resistance reading is higher or lower than the first reading.

   The placement of the ohmmeter's leads that results in the *lower* resistance reading is the arrangement that *forward biases* the diode. The lead that was connected to the diode's *anode* lead is the ohmmeter's *positive* (+) *lead*. Conversely, the lead that was connected to the diode's cathode is the *negative* (−) *lead*. In the following steps, we will refer to the ohmmeter's positive and negative leads as determined by this method, *regardless of how the leads are actually labeled on the ohmmeter*. It is very important to have this new convention in mind so that you will be able to test a transistor properly in the following steps. Thus, the internal circuit of the ohmmeter can be represented by the circuit of Figure 8–1.

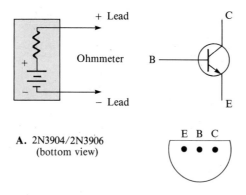

A. 2N3904/2N3906
   (bottom view)

FIGURE 8–1 *Schematic diagram of circuit.*

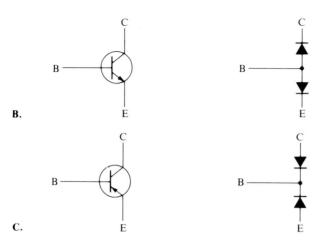

B.

C.

FIGURE 8–2 *Pin configuration of 2N3904 and 2N3906 transistors.*

If a DMM having a "diode check" feature is used, the display usually indicates the voltage drop across a good diode from anode to cathode when it is forward biased. When reverse biased, the DMM generally indicates some form of out-of-range condition, such as a blinking display or the letters "OL."

2. Using a 2N3904 npn transistor, whose schematic diagram and diode junction representation are shown in Figure 8–2B, connect the ohmmeter's positive lead to the transistor's base lead, with the ohmmeter's negative lead connected to the transistor's emitter lead. In this manner, you have *forward biased* the transistor's base-emitter diode junction.

Note whether the reading is at the high end or the low end of the meter's scale. Record your result in Table 8–1 as either "high" or "low."

3. Reverse the meter's leads so that the positive lead is connected to the emitter and the negative lead is connected to the base. Note the meter reading, and record either a "high" or a "low" result in Table 8–1.

4. Now connect the meter's positive lead to the base and the negative lead to the transistor's collector lead. Note whether the reading is at the high end or the low end of the meter's scale. Record your result in Table 8–1 as either "high" or "low."

5. Reverse the meter's leads so that the positive lead is connected to the collector and the negative lead is connected to the base. Note the meter reading, and record either a "high" or a "low" result in Table 8–1.

6. The base-emitter diode junction was forward biased in Step 2 and reverse biased in Step 3. If this junction is good, then you should have obtained a low reading in Step 2 and a high reading in Step 3.

81

The base-collector diode junction was forward biased in Step 4 and reverse biased in Step 5. If this junction is good, then you should have obtained a low reading in Step 4 and a high reading in Step 5. By examining the diode junction representation for an npn transistor shown in Figure 8–2B, you should be able to understand the operation of Steps 2 through 5.

7. Now connect the meter's positive lead to the collector and the negative lead to the transistor's emitter lead. Note the ohmmeter reading, and record this value in Table 8–1. Now reverse the meter's leads so that the positive lead is connected to the emitter and the negative lead is connected to the collector. Note the meter reading, and record this value in Table 8–1.

If the transistor is good, both readings should be virtually the same, namely, an infinite resistance. As neither of the two diodes between the collector and emitter leads is forward or reverse biased simultaneously, the result is basically an *open circuit*. Such a test of the collector-emitter junction is used to detect what is sometimes referred to as a "puncture short," which would result in a low-resistance path between the collector and emitter leads.

8. Using a 2N3906 pnp transistor, whose schematic diagram and diode junction representation are shown in Figure 8–2C, connect the ohmmeter's positive lead to the transistor's base lead, with the ohmmeter's negative lead connected to the transistor's emitter lead. Note whether the reading is at the high end or the low end of the meter's scale. Record your result in Table 8–2 as either "high" or "low."

9. Reverse the meter's leads so that the positive lead is connected to the emitter and the negative lead is connected to the base. Note the meter reading, and record either a "high" or a "low" result in Table 8–2.

10. Now connect the meter's positive lead to the base and the negative lead to the transistor's collector lead. Note whether the reading is at the high end or the low end of the meter's scale. Record your result in Table 8–2 as either "high" or "low."

11. Reverse the meter's leads so that the positive lead is connected to the collector and the negative lead is connected to the base. Note the meter reading, and record either a "high" or a "low" result in Table 8–2.

12. As in Step 7, connect the meter's positive lead to the collector and the negative lead to the transistor's emitter lead. Note the ohmmeter reading, and record this value in Table 8–2. Now reverse the meter's leads so that the positive lead is connected to the emitter and the negative lead is connected to the collector. Note the meter reading, and record this value in Table 8–2.

If the transistor is good, both readings should be virtually the same, namely, an infinite resistance. As neither of the two diodes between the collector and emitter leads is forward or reverse biased simultaneously, the result is basically an *open circuit.*

13. Compare the results of Tables 8–1 and 8–2. Note that if both transistors are good, npn and pnp transistors have opposite results. The base-emitter diode junction was forward biased in Step 9 and reverse biased in Step 8. If this junction is good, then you should have obtained a low reading in Step 9 and a high reading in Step 8.

The base-collector diode junction was forward biased in Step 11 and reverse biased in Step 10. If this junction is good, then you should have obtained a low reading in Step 11 and a high reading in Step 10. By examining the diode junction representation for a pnp transistor shown in Figure 8–2C, you should be able to understand the operation of Steps 8 through 12.

WHAT YOU HAVE DONE

This experiment demonstrated how to properly test npn and pnp transistors using either a VOM or DMM. This is because a transistor can be represented internally by two diode junctions.

NOTES

Name ______________________________ Date ____________

USING AN OHMMETER TO TEST TRANSISTOR DIODE JUNCTIONS

OBJECTIVES/PURPOSE:

SCHEMATIC DIAGRAM:

DATA FOR EXPERIMENT 8

TABLE 8–1 *2N3904 NPN transistor.*

Step Number	Ohmmeter Leads		Result
	+	**−**	
2	Base	Emitter	
3	Emitter	Base	
4	Base	Collector	
5	Collector	Base	
7	Collector	Emitter	
7	Emitter	Collector	

TABLE 8–2 *2N3906 PNP transistor.*

Step Number	Ohmmeter Leads		Result
	+	**−**	
8	Base	Emitter	
9	Emitter	Base	
10	Base	Collector	
11	Collector	Base	
12	Collector	Emitter	
12	Emitter	Collector	

Name _____ Date _____

RESULTS AND CONCLUSIONS:

REVIEW QUESTIONS FOR EXPERIMENT 8

1. An ohmmeter reads a low resistance when its negative lead is connected to a pnp transistor's base lead, with the meter's positive lead connected to the collector. The transistor junction is then
   (a) forward biased   (b) reverse biased
   (c) open   (d) shorted   (   )
2. An ohmmeter reads an infinite resistance when its positive lead is connected to an npn transistor's base lead, with the meter's negative lead connected to the emitter. The transistor junction is then
   (a) forward biased   (b) reverse biased
   (c) open   (d) shorted   (   )
3. An ohmmeter reads zero resistance when its negative lead is connected to an npn transistor's base lead, with the meter's positive lead connected to the collector. The transistor junction is then
   (a) forward biased   (b) reverse biased
   (c) open   (d) shorted   (   )
4. An ohmmeter reads a high resistance when its positive lead is connected to a transistor's base lead, with the meter's negative lead connected to the emitter or the collector leads. The transistor is then
   (a) forward biased   (b) reverse biased
   (c) an npn type   (d) a pnp type   (   )

NOTES

9

TRANSISTOR BASE BIASING

PURPOSE AND BACKGROUND

The purpose of this experiment is to verify the voltages and currents in a base-biased circuit as well as to construct its dc load line. In spite of its simplicity, a base-biased circuit does not effectively stabilize a transistor's quiescent point. Consequently, the Q point is affected by the transistor's current gain (β).

Text References: 5–1, The DC Operating Point; 5–2, Base Bias.

REQUIRED PARTS AND EQUIPMENT

Resistors (1/4 W):
- ☐ 1 kΩ
- ☐ 560 kΩ
- ☐ 1 MΩ potentiometer
- ☐ Two 2N3904 npn silicon transistors

- ☐ 0–15 V dc power supply
- ☐ VOM or DMM (preferred)
- ☐ Breadboarding socket

USEFUL FORMULAS

Quiescent dc base voltage

$$(1)\ V_B = V_{CC} - I_B R_B = V_{BE}$$

Quiescent dc collector (emitter) current

$$(2)\ I_C \simeq \frac{V_{CC} - V_{BE}}{R_B/\beta}$$

Quiescent dc base current

$$(3)\ I_B = \frac{V_{CC} - V_{BE}}{R_B}$$

Quiescent dc collector-to-emitter voltage

$$(4)\ V_{CE} = V_{CC} - I_C R_C$$

dc load line

$$(5)\ I_{C(\text{sat})} \cong \frac{V_{CC}}{R_C} \qquad (\text{saturation})$$

$$(6)\ V_{CE(\text{off})} = V_{CC} \qquad (\text{cutoff})$$

In general, make

$$(7)\ V_{CC} \gg V_{BE}$$

PROCEDURE

1. Wire the circuit shown in the schematic diagram of Figure 9–1, and apply power to the breadboard.
2. With your VOM or DMM, measure the voltage across the base and collector resistors, and, using Ohm's law, determine the corresponding currents, recording your values in Table 9–1. From these two sets of values, determine the dc current gain or beta (β_{dc}) for this transistor so that

$$\beta_{\text{dc}} = \frac{I_C}{I_B}$$

   Record this value of beta in Table 9–1.
3. Use your VOM or DMM to measure individually V_B and V_{CE}. Record your results in Table 9–1.
4. Compare the values of Step 3 with the expected values, using the value of β_{dc} determined in Step 2 and a typical base-emitter voltage of 0.7 V. Record these values in Table 9–1.
5. Now use a hand-held hair dryer to blow hot air against the transistor's case for a few seconds while measuring the

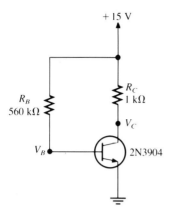

FIGURE 9–1 *Schematic diagram of circuit.*

collector current using your VOM or DMM. Does the collector current increase or decrease?

You should find that the collector current increases, which in turn causes the circuit's Q point to change.

6. Using Equations 5 and 6 in the "Useful Formulas" section of this experiment, determine the saturation and cutoff points on the dc load line for this circuit, and record these values in Table 9–2. On the blank graph provided, plot the dc load line, using the calculated values of $I_{C(\text{sat})}$ and $V_{CE(\text{off})}$ as the endpoints of the load line. Now plot the Q point based on the measured values of I_C and V_{CE} on the same graph. What do you notice about the Q point?

You should find that the measured Q point lies essentially on the dc load line.

7. Using a different 2N3904 transistor, repeat Steps 2 through 5, and record your results in Table 9–1. Do you find any differences between the two transistors?

You will usually find that the two transistors give different values for the quiescent voltages and currents. In addition, you will usually find differences in the dc current gains.

8. Disconnect the power from the breadboard and replace the 560-kΩ resistor (R_B) with a 1-MΩ potentiometer. Again apply power to the breadboard and connect a voltmeter between the transistor's collector terminal and ground.

9. Now vary the resistance of the potentiometer until V_{CE} as read by the voltmeter reaches a minimum value, $V_{CE(\text{sat})}$. Then measure the corresponding collector current, $I_{C(\text{sat})}$. Record both values in Table 9–2.

10. Continue to vary the resistance of the 1-MΩ potentiometer until V_{CE} reaches a maximum value, $V_{CE(\text{off})}$. Then measure the

corresponding collector current $I_{C(off)}$. If the collector current is not essentially zero, then temporarily disconnect one lead of the potentiometer from the circuit so that the base current is zero. The collector current should also be zero. Measure the corresponding collector-emitter voltage, $V_{CE(off)}$. Record both $I_{C(off)}$ and $V_{CE(off)}$ in Table 9–2.

At saturation, $V_{CE(sat)}$ is ideally zero, while at cutoff, $I_{C(off)}$ is zero. Plot the values for I_C and V_{CE} at cutoff and saturation on the graph constructed in Step 6. You should find that both points lie essentially on the dc load line very close to the ideal endpoints of cutoff and saturation.

11. If you disconnected the potentiometer in Step 10, reconnect the potentiometer as in Step 8. Vary the potentiometer so that you are able to measure about five combinations of I_C and V_{CE} over the active region of the dc loan line, recording all values in Table 9–2. Then plot these values on the graph. As in Step 10, each point should lie essentially on the dc load line, as the load line is a plot of all possible combinations of I_C and V_{CE}.

WHAT YOU HAVE DONE

This experiment verified the voltages and currents in a base-biased circuit as well as constructing the dc load line for the circuit. In addition, the effect of temperature on the stability of the bias circuit was also demonstrated.

Name ______________________ Date ____________

TRANSISTOR BASE BIASING

OBJECTIVES/PURPOSE:

SCHEMATIC DIAGRAM:

Name _____   Date _____

DATA FOR EXPERIMENT 9

TABLE 9–1

Parameter	Transistor 1		Transistor 2	
	Measured Value	Expected Value	Measured Value	Expected Value
I_B				
I_C				
β_{dc}		███████		███████
V_B		0.7 V (typical)		0.7 V (typical)
V_{CE}				

TABLE 9–2

Condition	Calculated Values		Measured Values	
	I_C	V_{CE}	I_C	V_{CE}
Saturation (Step 9)				
Cutoff (Step 10)				
Active Region (Step 11)				

Name _____ Date _____

DATA FOR EXPERIMENT 9

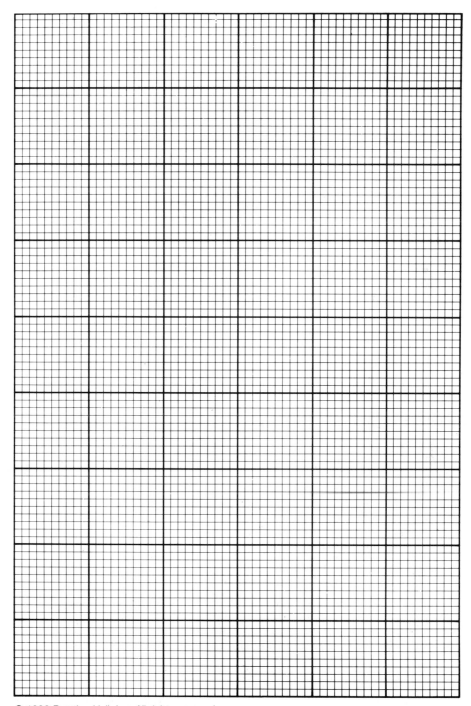

Name _____ Date _____

RESULTS AND CONCLUSIONS:

REVIEW QUESTIONS FOR EXPERIMENT 9

1. For the circuit of Figure 9–1, if $\beta = 150$, then I_B is
   (a) 10 μA   **(b)** 15 μA
   (c) 20 μA   **(d)** 25 μA   (   )
2. If β of the transistor in the circuit of Figure 9–1 increases, then
   (a) I_B decreases   **(b)** I_C increases
   (c) V_{CE} decreases   **(d)** all of the above   (   )
3. If R_B is made smaller in the circuit of Figure 9-1, then
   (a) I_B decreases   **(b)** I_C increases
   (c) V_{CE} decreases   **(d)** all of the above   (   )
4. The collector saturation current for the circuit of Figure 9–1 is
   approximately
   (a) 4 mA   **(b)** 6 mA   **(c)** 10 mA   **(d)** 15 mA   (   )
5. At cutoff, the collector-to-emitter voltage for the circuit of Figure
   9–1 is
   (a) 5 V   **(b)** 7.5 V   **(c)** 10 V   **(d)** 15 V   (   )

96

10

TRANSISTOR EMITTER BIASING

PURPOSE AND BACKGROUND

The purpose of this experiment is to verify the voltages and currents in an emitter-biased circuit as well as to construct its dc load line. Unlike other biasing schemes, emitter bias uses both a positive and a negative supply voltage. In this manner, the base is approximately at ground while the negative emitter supply voltage forward biases the base-emitter junction.

   Text References: 5–1, The DC Operating Point; 5–3, Emitter Bias.

REQUIRED PARTS AND EQUIPMENT

Resistors (1/4 W):
   ☐ Two 1 kΩ
   ☐ 4.7 kΩ
☐ Two 2N3904 npn silicon transistors

☐ Two 0–15 V dc power supplies
☐ VOM or DMM (preferred)
☐ Breadboarding socket

USEFUL FORMULAS

Quiescent dc emitter voltage

$$(1)\ V_E = V_{EE} - V_{BE}$$

Quiescent dc base voltage

$$(2)\ V_B = V_{EE} - I_E R_E - V_{BE}$$

Quiescent dc collector (emitter) current

$$(3)\ I_C \cong \frac{V_{EE} - V_{BE}}{R_E + (R_B/\beta)} \qquad (I_C \cong I_E \text{ for large } \beta)$$

Quiescent dc base current

$$(4)\ I_B = \frac{V_B}{R_B} \cong \frac{V_{EE} - V_{BE}}{\beta R_E + R_B}$$

Quiescent dc collector-to-emitter voltage

$$(5)\ V_{CE} = V_{CC} - I_C R_C + V_{BE}$$

dc load line

$$(6)\ I_{C(\text{sat})} = \frac{V_{CC} + V_{EE}}{R_C + R_E} \qquad (\text{saturation})$$

$$(7)\ V_{CE(\text{off})} = V_{CC} + V_{EE} \qquad (\text{cutoff})$$

In general, make

$$(8)\ R_E \gg \frac{R_B}{\beta}$$

$$(9)\ V_{EE} \gg V_{BE}$$

PROCEDURE

1. Wire the circuit shown in the schematic diagram of Figure 10–1, and apply power to the breadboard.
2. With your VOM or DMM, measure the base, emitter and collector voltages, with respect to ground, and measure the base and collector currents, recording your values in Table 10–1. From these two sets of values, determine the dc current gain, or beta (β), for this transistor so that

$$\beta_{\text{dc}} = \frac{I_C}{I_B}$$

Record this value of dc beta in Table 10–1.

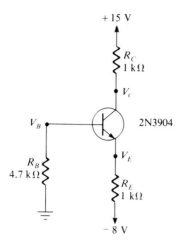

FIGURE 10–1 *Schematic diagram of circuit.*

3. Then measure V_{CE}, and record your results in Table 10–1.
4. Compare these values and those of Steps 2 and 3 with the expected values, using the value of beta determined in Step 2 and a typical base-emitter voltage of 0.7 V. Record these values in Table 10–1.
5. Now use a hand-held hair dryer to blow hot air against the transistor's case for a few seconds while measuring the emitter current using your VOM or DMM. Does the emitter current increase or decrease?

    You should find that the emitter current increases, which in turn causes the circuit's Q point to change.
6. Using Equations 6 and 7 in the "Useful Formulas" section of this experiment, determine the saturation and cutoff points on the dc load line for this circuit, and record these values in Table 10–2. On the blank graph provided, plot the dc load line, using the calculated values of $I_{C(sat)}$ and $V_{CE(off)}$ as the endpoints of the load line. Now plot the Q point based on the measured values of I_C and V_{CE} on the same graph. What do you notice about the Q point?

    You should find that the measured Q point lies essentially on the dc load line.
7. Using a different 2N3904 transistor, repeat Steps 2 through 5, and record your results in Table 10–1. Do you find any differences between the two transistors?

    You will usually find that the two transistors give different values for the quiescent voltages and currents. In addition, you will usually find differences in the current gains.

WHAT YOU HAVE DONE

This experiment verified the voltages and currents in an emitter-biased circuit using two power supplies as well as constructing the dc load line for the circuit. In addition, the effect of temperature on the stability of the bias circuit was also demonstrated.

Name _____ Date _____

TRANSISTOR EMITTER BIASING

OBJECTIVES/PURPOSE:

SCHEMATIC DIAGRAM:

DATA FOR EXPERIMENT 10

TABLE 10–1

Parameter	Transistor 1		Transistor 2	
	Measured Value	Expected Value	Measured Value	Expected Value
I_C				
I_B				
β_{dc}		■		■
V_B				
V_C				
V_E				
V_{CE}				

TABLE 10–2

Parameter	Calculated Value
$V_{CE(off)}$	V
$I_{C(sat)}$	mA

Name _____   Date _____

DATA FOR EXPERIMENT 10

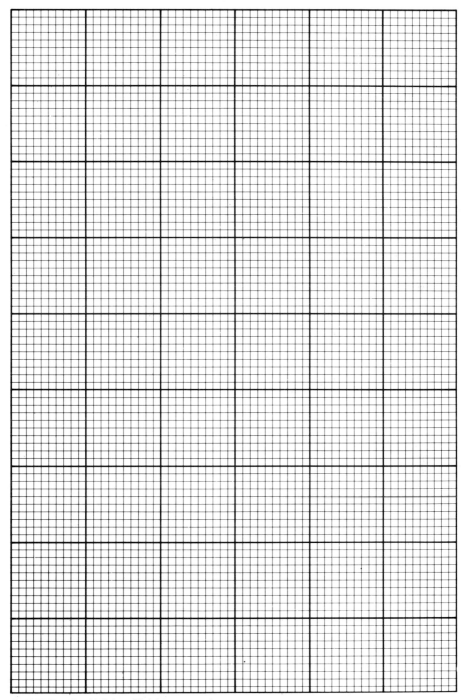

Name _____ Date _____

RESULTS AND CONCLUSIONS:

REVIEW QUESTIONS FOR EXPERIMENT 10

1. For the circuit of Figure 10–1, if $\beta = 100$, then V_C is
   (a) 6 V   (b) 8 V   (c) 10 V   (d) 15 V   (   )
2. If β increases for the transistor of Figure 10–1, then
   (a) I_C decreases   (b) V_B decreases
   (c) V_{CE} decreases   (d) all of the above   (   )
3. If R_B is made smaller in the circuit of Figure 10–1, then
   (a) I_B decreases   (b) I_C increases
   (c) V_{CE} increases   (d) all of the above   (   )
4. The collector saturation current for the circuit of Figure 10–1 is
   approximately
   (a) 4 mA   (b) 7.5 mA   (c) 11.5 mA   (d) 23 mA   (   )
5. At cutoff, the collector-to-emitter voltage for the circuit of Figure
   10–1 is
   (a) 6 V   (b) 8 V   (c) 15 V   (d) 23 V   (   )

104

11

TRANSISTOR VOLTAGE-DIVIDER BIASING

PURPOSE AND BACKGROUND

The purpose of this experiment is to verify the voltages and currents in a transistor voltage-divider bias circuit as well as to construct the dc load line. Voltage-divider bias is often used because the base current is made small compared to the currents through the two base ("voltage-divider") resistors. Consequently, the base voltage and therefore the collector current are stabilized against changes in the transistor beta.

Text References: 5–1, The DC Operating Point; 5–4, Voltage-Divider Bias.

REQUIRED PARTS AND EQUIPMENT

Resistors (1/4 W):
- ☐ Two 1 kΩ
- ☐ 4.7 kΩ
- ☐ 10 kΩ
- ☐ 10-kΩ potentiometer

- ☐ Two 2N3904 npn silicon transistors
- ☐ 0–15 V dc power supply
- ☐ VOM or DMM (preferred)
- ☐ Breadboarding socket

USEFUL FORMULAS

Quiescent dc base voltage

$$(1) \ V_B \cong \left(\frac{R_2}{R_1 + R_2} \right) V_{CC}$$

Quiescent dc emitter voltage

$$(2) \ V_E = V_B - V_{BE}$$

Quiescent dc collector (emitter) current

$$(3) \ I_C \simeq \frac{V_E}{R_E} \qquad (I_C \simeq I_E \text{ for large } \beta)$$

Quiescent collector voltage

$$(4) \ V_C = V_{CC} - I_C R_C$$

Quiescent dc collector-to-emitter voltage

$$(5) \ V_{CE} \cong V_{CC} - I_C(R_C + R_E) = V_C - V_E$$

dc load line

$$(6) \ I_{c(\text{sat})} \cong \frac{V_{CC}}{R_C + R_E} \qquad (\text{saturation})$$

$$(7) \ V_{CE(\text{off})} = V_{CC} \qquad (\text{cutoff})$$

In general, make

$$(8) \ R_1 \parallel R_2 \ll \beta R_E$$

PROCEDURE

1. Using a typical value for the base-emitter voltage of a silicon transistor (0.7 V), calculate the expected values of the quiescent dc base voltage (V_B), emitter voltage (V_E), collector voltage (V_C), and collector-emitter voltage (V_{CE}) for the voltage-divider bias circuit shown in the schematic diagram of Figure 11–1. Record these values in Table 11–1.
2. Now wire the circuit shown in the schematic diagram of Figure 11–1, and apply power to the breadboard.
3. Use your VOM or DMM to measure in turn V_B, V_C, V_E, and V_{CE}. Record your results in Table 11–1, comparing these measured values with the expected voltages determined in Step 1. Your results should agree within 10 percent.
4. Now measure the quiescent collector current and compare this value with the expected value (Equation 3). Record this value in Table 11–1.

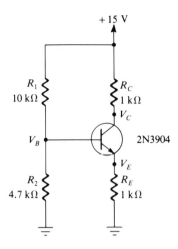

FIGURE 11–1 *Schematic diagram of circuit.*

5. Now use a hand-held hair dryer to blow hot air against the transistor's case for a few seconds while measuring the collector current using your VOM or DMM. Does the collector current increase or decrease?

   You should find that there is very little change in the collector current, or that the change is not as fast or as great as is obtained with a circuit using base biasing. The well-designed voltage-divider bias circuit makes the Q point essentially independent of transistor beta.

6. Using Equations 6 and 7 in the "Useful Formulas" section of this experiment, determine the saturation and cutoff points on the dc load line for this circuit, and record these values in Table 11–2. On the graph provided, plot the dc load line, using the calculated values of $I_{C(sat)}$ and $V_{CE(off)}$ as the endpoints of the load line. On the same graph, plot the Q point based on the measured values of I_C and V_{CE}. What do you notice about the Q point?

   You should find that the measured Q point lies essentially on the dc load line.

7. Using a different 2N3904 transistor, repeat Steps 3 through 6, recording your results in Table 11–1. Do you find any differences between the two transistors?

   You should find that there are essentially no differences in the measured values, as the voltage-divider bias circuit basically makes the circuit's quiescent voltages and current independent of beta.

8. Using transistor #2, disconnect power from the breadboard and replace resistors R_1 and R_2 with a 10-kΩ potentiometer, as shown in Figure 11–2.

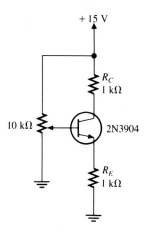

FIGURE 11–2 *Schematic diagram from Step 8.*

9. Connect the power to the breadboard and connect a voltmeter between the transistor's collector and emitter terminals. Slowly vary the 10-kΩ potentiometer until V_{CE} as read by the voltmeter reaches a minimum value, $V_{CE(\text{sat})}$. Then measure the corresponding collector current, $I_{C(\text{sat})}$. Record both values in Table 11-2.

10. Continue to vary the resistance of the 10-kΩ potentiometer until V_{CE} reaches a maximum value, $V_{CE(\text{off})}$. Then measure the corresponding collector current, $I_{C(\text{off})}$. Record both $I_{C(\text{off})}$ and $V_{CE(\text{off})}$ in Table 11–2.

   At saturation, $V_{CE(\text{sat})}$ is ideally zero, while at cutoff, $I_{C(\text{off})}$ is zero. Plot the values for I_C and V_{CE} at cutoff and saturation on the graph constructed in Step 6. You should find that both points lie essentially on the dc load line very close to the ideal endpoints of cutoff and saturation.

11. Vary the potentiometer so that you are able to measure about five combinations of I_C and V_{CE} over the active region of the dc load line, recording all values in Table 11–2. Then plot these values on the graph. As in Step 10, each point should lie essentially on the dc load line, as the load line is a plot of all possible combinations of I_C and V_{CE}.

WHAT YOU HAVE DONE

This experiment verified the voltages and currents in a voltage divider-biased circuit as well as constructing the dc load line for the circuit. In addition, the effect of temperature on the stability of the bias circuit was also demonstrated.

Name ______________________________ Date ____________

TRANSISTOR VOLTAGE DIVIDER BIASING

OBJECTIVES/PURPOSE:

SCHEMATIC DIAGRAM:

DATA FOR EXPERIMENT 11

TABLE 11–1

Parameter	Measured Values		Expected Value
	Transistor 1	Transistor 2	
V_B			
V_C			
V_E			
V_{CE}			
I_C			

TABLE 11–2

Condition	Calculated Values		Measured Values	
	I_C	V_{CE}	I_C	V_{CE}
Saturation (Step 9)				
Cutoff (Step 10)				
Active Region (Step 11)				

Name _____   Date _____

DATA FOR EXPERIMENT 11

Name _____ Date _____

RESULTS AND CONCLUSIONS:

REVIEW QUESTIONS FOR EXPERIMENT 11

1. For the circuit of Figure 11–1, I_E is approximately
   (a) 2 mA   (b) 4 mA   (c) 6 mA   (d) 7.5 mA   (   )
2. For the transistor in the circuit of Figure 11–1, if β increases,
   V_B will
   (a) decrease   (b) increase
   (c) remain essentially the same   (   )
3. If R_2 is increased, then
   (a) V_B decreases   (b) I_C decreases
   (c) V_{CE} increases   (d) V_C decreases   (   )
4. The collector saturation current for the circuit of Figure 11–1 is
   approximately
   (a) 4 mA   (b) 7.5 mA   (c) 10 mA   (d) 15 mA   (   )
5. At cutoff, the collector-to-emitter voltage for the circuit of Figure
   11–1 is
   (a) 4 V   (b) 8 V   (c) 10 V   (d) 15 V   (   )

112

12

TRANSISTOR COLLECTOR-
FEEDBACK BIASING

PURPOSE AND BACKGROUND

The purpose of this experiment is to verify the voltages and currents in a collector-feedback bias circuit as well as to construct its dc load line. This arrangement is different from other biasing schemes in that the collector voltage provides the bias for the base-emitter junction. The result is a very stable Q point, which reduces the effects of transistor beta.

Text References: 5–1, The DC Operating Point; 5–5 Collector-Feedback Bias.

REQUIRED PARTS AND EQUIPMENT

Resistors (1/4 W):
  ☐ 2.7 kΩ
  ☐ 560 kΩ
☐ Two 2N3904 npn silicon
  transistors

☐ 0–15 V dc power supply
☐ VOM or DMM (preferred)
☐ Breadboarding socket

USEFUL FORMULAS

Quiescent dc base voltage

    (1) $V_B = V_{BE}$

Quiescent dc collector (emitter) current

    (2) $I_C \simeq \dfrac{V_{CC} - V_{BE}}{R_C + (R_B/\beta)}$

Quiescent dc base current

    (3) $I_B \simeq \dfrac{V_{CC} - V_{BE}}{\beta R_C + R_B}$

Quiescent dc collector-to-emitter voltage

    (4) $V_{CE} \simeq V_{CC} - I_C R_C$

dc load line

    (5) $I_{C(\text{sat})} \cong \dfrac{V_{CC}}{R_C}$ (saturation)

    (6) $V_{CE(\text{off})} = V_{CC}$ (cutoff)

In general, make

    (7) $R_C \gg \dfrac{R_B}{\beta}$

    (8) $V_{CC} \gg V_{BE}$

PROCEDURE

1. Wire the circuit shown in the schematic diagram of Figure 12–1, and apply power to the breadboard.
2. With your VOM or DMM and Ohm's law, measure the quiescent base and collector currents, recording your values in Table 12–1. From these two values, determine the dc current gain or dc beta (β_{dc}) for this transistor so that

    $\beta_{\text{dc}} = \dfrac{I_C}{I_B}$

Record this value of beta in Table 12–1.
3. Use your VOM or DMM to measure individually V_B and V_{CE}. Record your results in Table 12–1.
4. Using the value for dc beta determined in Step 2 for both transistors, and using a typical base-emitter voltage of 0.7 V, calculate the expected values for I_C, I_B, and V_{CE} (Equations 2, 3, and 4). Record your results in Table 12–1. The measured results should be within approximately 10 percent of the expected results.

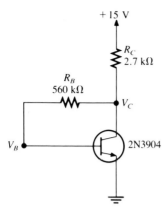

+ 15 V

R_C
2.7 kΩ

R_B
560 kΩ

V_C

V_B

2N3904

FIGURE 12–1 *Schematic diagram of circuit.*

5. Now use a hand-held hair dryer to blow hot air against the transistor's case for a few seconds while measuring the collector current using your VOM or DMM. Does the collector current increase or decrease?

   You should find that there is very little change in the collector current, or that the change is not as fast or as great as is obtained with a circuit using base biasing. A well-designed collector-feedback bias circuit makes the Q point essentially independent of transistor beta.

6. Using Equations 5 and 6 in the "Useful Formulas" section of this experiment, determine the saturation and cutoff points on the dc load line for this circuit, and record these values in Table 12–2. On the blank graph provided, plot the dc load line, using the calculated values of $I_{C(\text{sat})}$ and $V_{CE(\text{off})}$ as the endpoints of the load line. On the same graph, plot the Q point based on the measured values of I_C and V_{CE}. What do you notice about the Q point?

   You should find that the measured Q point lies essentially on the dc load line.

7. Using a different 2N3904 transistor, repeat Steps 2 through 5, and record your results in Tables 12–1 and 12–2. Do you find any differences between the two transistors?

WHAT YOU HAVE DONE

This experiment verified the voltages and currents in a collector-feedback biased circuit as well as constructing the dc load line for the circuit. In addition, the effect of temperature on the stability of the bias circuit was also demonstrated.

115

NOTES

Name ______________________ Date __________

TRANSISTOR COLLECTOR-FEEDBACK BIASING

OBJECTIVES/PURPOSE:

SCHEMATIC DIAGRAM:

DATA FOR EXPERIMENT 12

TABLE 12–1

Parameter	Transistor 1		Transistor 2	
	Measured	Expected	Measured	Expected
I_B				
I_C				
β_{dc}		■		■
V_B		0.7 V (typical)		0.7 V (typical)
V_{CE}				

TABLE 12–2

Parameter	Expected Value
$I_{C(sat)}$	mA
$V_{CE(off)}$	V

Name _____ Date _____

DATA FOR EXPERIMENT 12

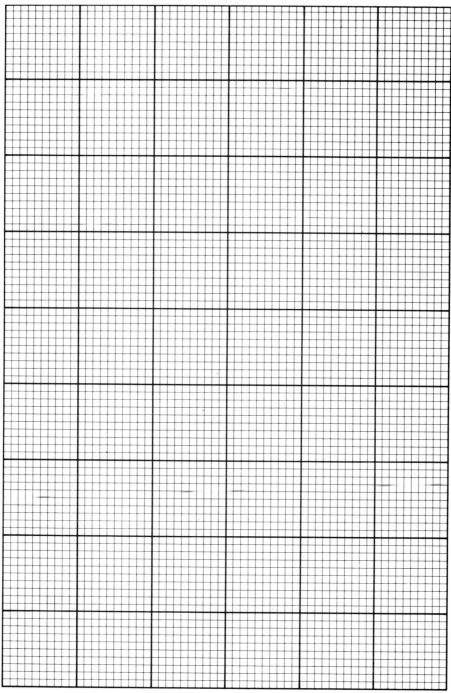

Name _____   Date _____

RESULTS AND CONCLUSIONS:

REVIEW QUESTIONS FOR EXPERIMENT 12

1. For the circuit of Figure 12–1, if $\beta = 200$, then I_C is approximately
   (a) 2.2 mA   **(b)** 3.9 mA
   (c) 4.8 mA   **(d)** 14.3 mA   (   )

2. In the circuit of Figure 12–1, if β of the transistor increases, then
   (a) I_B decreases   **(b)** I_C increases   **(c)** V_{CE} increases
   (d) all of the above   (   )

3. The collector saturation current for the circuit of Figure 12–1 is approximately
   (a) 4 mA   **(b)** 6 mA   **(c)** 10 mA   **(d)** 15 mA   (   )

4. At cutoff, the collector-to-emitter voltage for the circuit of Figure 12–1 is
   (a) 5 V   **(b)** 7.5 V   **(c)** 10 V   **(d)** 15 V   (   )

120

13

THE COMMON-EMITTER AMPLIFIER

PURPOSE AND BACKGROUND

The purposes of this experiment are to (1) demonstrate the operation and characteristics of the small-signal common-emitter amplifier and (2) investigate what influences its voltage gain. The common-emitter amplifier is characterized by application of the amplifier input signal to the base lead while its output is taken from the collector, which always gives a 180° phase shift.

Text Reference: 6–3, Common-Emitter Amplifiers.

REQUIRED PARTS AND EQUIPMENT

Resistors (1/4 (W)):
- ☐ 150 Ω
- ☐ 2.7 kΩ
- ☐ Two 3.9 kΩ
- ☐ 4.7 kΩ
- ☐ 10 kΩ

Capacitors (25 V):
- ☐ Two 2.2 μF
- ☐ 10 μF

- ☐ 2N3904 npn silicon transistor
- ☐ 0–15 V dc power supply
- ☐ Signal generator
- ☐ VOM or DMM (preferred)
- ☐ Dual trace oscilloscope
- ☐ Breadboarding socket

USEFUL FORMULAS

Voltage gain from base to collector

$$(1) \quad A_v = \frac{v_{out}}{v_{in}}$$

$$(2) \quad A_v = \frac{R_c \parallel R_L}{R_{E1} + r_e} \qquad \text{(normal circuit)}$$

$$(3) \quad A_v = \frac{R_C}{R_{E1} + r_e} \qquad \text{(no load)}$$

$$(4) \quad A_v = \frac{R_C \parallel R_L}{R_{E1} + R_{E2} + r_e} \qquad \text{(no bypass capacitor)}$$

Transistor ac emitter resistance (at normal room temperature)

$$(5) \quad r_e \cong \frac{25 \text{ mV}}{I_E}$$

Quiescent dc base voltage

$$(6) \quad V_B = \left(\frac{R_2}{R_1 + R_2}\right) V_{CC}$$

Quiescent dc emitter voltage

$$(7) \quad V_E = V_B - V_{BE}$$

Quiescent dc emitter current

$$(8) \quad I_E = \frac{V_E}{R_{E1} + R_{E2}} \qquad (I_C \simeq I_E \text{ for large } \beta)$$

Quiescent dc collector voltage

$$(9) \quad V_C = V_{CC} - I_C R_C$$

Quiescent dc collector-emitter voltage

$$(10) \quad V_{CE} = V_{CC} - I_C(R_C + R_{E1} + R_{E2})$$

PROCEDURE

1. Wire the circuit shown in Figure 13–1, omitting the signal generator and the power supply.
2. After you have checked all connections, apply the 15-V supply voltage the breadboard. With a VOM or DMM, individually measure the transistor dc base, emitter, and collector voltages with respect to ground, recording your results in Table 13–1. Based on the resistor values of Figure 13–1, determine the expected values of these three voltages (Equations 6, 7, and 9), assuming a base-emitter voltage drop of 0.7 V, and compare them with the measured values in Table 13–1.

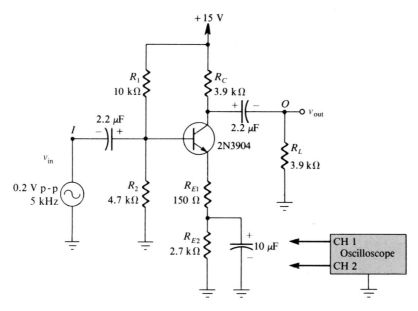

FIGURE 13–1 *Schematic diagram of circuit.*

3. Connect Channel 1 of your oscilloscope to point I (v_{in}) and Channel 2 to point O (v_{out}). Then connect the signal generator to the circuit as shown in Figure 13–1, and adjust the sine wave output level of the generator at 0.2 V *peak-to-peak* at a frequency of 5 kHz.

   You should observe that the output signal level (v_{out}) is *greater* than the input level (v_{in}). In addition, v_{out} is inverted, or 180° out-of-phase, with respect to the input. These points are two major characteristics of a common-emitter amplifier. In order to observe the phase shift, you must display both signals *simultaneously* on the oscilloscope; otherwise you will not see any phase shift.

4. Using the *measured* value for the dc emitter voltage obtained in Step 2, calculate the dc emitter current (Equation 8) and the resultant transistor ac emitter resistance, r_e (Equation 5). Record these values in Table 13–2.

5. With an oscilloscope, measure the ac peak-to-peak voltage at the junction of R_{E1} (150 Ω) and R_{E2} (2.7 kΩ). Even at the oscilloscope's highest input sensitivity, you should measure virtually no ac voltage at this point. The 10-μF bypass capacitor, in parallel with R_{E2}, serves essentially as a short-circuit path to ground since its reactance at 5 kHz is very small compared with the 2.7-kΩ resistance. Consequently, the junction of R_{E1} and R_{E2} is effectively at *ac ground*.

   Measure the ac peak-to-peak voltage at the transistor's emitter lead. Note that the ac voltage is slightly less than the input

123

voltage, v_{in}. In addition, both signals are *in-phase*. Since R_{E1} is used to minimize the temperature effects of r_e, an ac voltage will be present at the emitter terminal, which is approximately equal to, and in-phase with, the input signal.

6. Calculate the expected voltage gain from base to collector using Equation 2 given in the "Useful Formulas" section of this experiment and record the value in Table 13–3. Now measure the actual voltage gain by dividing the peak-to-peak output voltage v_{out} by the peak-to-peak input voltage v_{in}, recording your result in Table 13–3.

7. Now remove R_L. You should observe that the output voltage level increases. It does so because the load resistance affects the voltage gain of the amplifier stage. As in Step 6, experimentally determine the voltage gain by measuring v_{out} and v_{in}, comparing your measured result with the expected value (Equation 3). Record your results in Table 13–3.

8. Reconnect the 3.9-kΩ load resistor as in the original circuit of Figure 13–1. Remove the 10-μF emitter bypass capacitor from the circuit. You should observe that the output voltage decreases tremendously. It does so because the total ac emitter resistance is now $R_{E1} \mid R_{E2}$, in addition to the transistor's internal ac emitter resistance, r_e. As in the previous two steps, experimentally determine the voltage gain by measuring v_{out} and v_{in}, comparing your results with the expected value (Equation 4). Record your results in Table 13–3.

From the results in Table 13–3, you should now understand how both the emitter bypass capacitor and the load resistance affect the base-to-collector voltage gain of a common-emitter amplifier.

WHAT YOU HAVE DONE

This experiment demonstrated the operation and characteristics of a small-signal common-emitter amplifier, which has a 180° phase shift. Here, the input signal is applied to the transistor's base lead, while the output signal is taken from the collector lead. The experiment also showed how the load resistance and emitter bypass capacitor each influence the circuit's voltage gain.

Name ______________________ Date ____________

THE COMMON-EMITTER AMPLIFIER

OBJECTIVES/PURPOSE:

SCHEMATIC DIAGRAM:

Name _____ Date _____

DATA FOR EXPERIMENT 13

TABLE 13–1

Parameter	Measured Value	Expected Value	% Error
V_B			
V_E			
V_C			

TABLE 13–2

Parameter	Value
I_E (calculated)	
r_e (calculated)	

TABLE 13–3

Condition	V_{in}	V_{out}	Measured Gain	Expected Gain	% Error
Normal circuit (Step 6)					
No load (Step 7)					
No bypass capacitor (Step 8)					

Name _____ Date _____

RESULTS AND CONCLUSIONS:

REVIEW QUESTIONS FOR EXPERIMENT 13

1. For the circuit of Figure 13–1, the voltage gain from base to collector is approximately
   (a) 1   **(b)** 12   **(c)** 112   **(d)** 224   (   )
2. The output signal of a common-emitter amplifier is out-of-phase with the input by
   (a) 0°   **(b)** 45°   **(c)** 90°   **(d)** 180°   (   )
3. If the emitter bypass capacitor in Figure 13–1 is removed, the amplifier voltage gain will
   (a) increase   **(b)** decrease
   (c) remain essentially the same   (   )
4. If the emitter bypass capacitor in Figure 13–1 is shorted,
   (a) the voltage gain will increase
   (b) the voltage gain will remain the same
   (c) the transistor will saturate
   (d) none of the above   (   )
5. If the load resistor R_L in the circuit of Figure 13–1 is made larger, the amplifier voltage gain will
   (a) increase   **(b)** decrease
   (c) remain essentially the same   (   )

NOTES

14

THE COMMON-COLLECTOR AMPLIFIER (EMITTER-FOLLOWER)

PURPOSE AND BACKGROUND

The purposes of this experiment are (1) to demonstrate the operation and characteristics of the small-signal common-collector amplifier and (2) to investigate what influences its voltage gain. The common-collector amplifier, often referred to as an *emitter-follower*, is characterized by application of the amplifier input signal to the base lead while its output is taken from the emitter. The output signal is never larger than the input but is always in-phase with the input. Consequently, the output *follows* the input. The main advantage is that the input impedance of a common-collector amplifier is generally much higher than for other bipolar transistor circuits.

Text Reference: 6–4, Common-Collector Amplifiers.

REQUIRED PARTS AND EQUIPMENT

Resistors (1/4 W):
- [ ] 68 Ω
- [ ] 100 Ω
- [ ] Two 1 kΩ
- [ ] 22 kΩ
- [ ] 27 kΩ

Capacitors (25 V):
- [ ] 2.2 μF
- [ ] 100 μF

- [ ] 2N3904 npn silicon transistor
- [ ] 0–15 V dc power supply
- [ ] signal generator
- [ ] VOM or DMM (preferred)
- [ ] Dual trace oscilloscope
- [ ] Breadboarding socket

USEFUL FORMULAS

Voltage gain from base to emitter

$$(1) \quad A_v = \frac{v_{\text{out}}}{v_{\text{in}}}$$

$$(2) \quad A_v = \frac{R_E \parallel R_L}{(R_E \parallel R_L) + r_e}$$

Transistor ac emitter resistance (at normal room temperature)

$$(3) \quad r_e \cong \frac{25 \text{ mV}}{I_E}$$

Quiescent dc base voltage

$$(4) \quad V_B = \left(\frac{R_2}{R_1 + R_2}\right) V_{CC}$$

Quiescent dc emitter voltage

$$(5) \quad V_E = V_B - V_{BE}$$

Quiescent dc emitter current

$$(6) \quad I_E = \frac{V_E}{R_E}$$

Amplifier input impedance

$$(7) \quad R_{\text{in}} = R_1 \parallel R_2 \parallel \beta[(R_E \parallel R_L) + r_e]$$

PROCEDURE

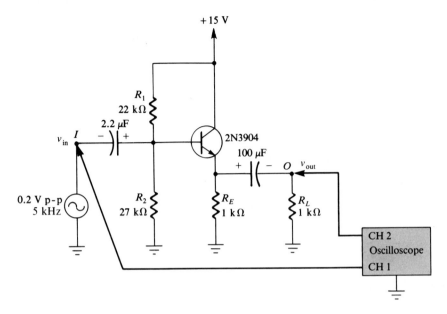

FIGURE 14–1 *Schematic diagram of circuit.*

1. Wire the circuit shown in Figure 14–1, omitting the signal generator and the power supply.
2. After you have checked all connections, apply the 15-V supply voltage to the breadboard. With a VOM or DMM, individually measure the transistor dc base and emitter voltages with respect to ground, recording your results in Table 14–1. Based on the resistor values of Figure 14–1, determine the expected values of these two voltages (Equations 4 and 5), assuming a base-emitter voltage drop of 0.7 V, and compare them with the measured values in Table 14–1.
3. Connect Channel 1 of your oscilloscope at point I (v_{in}) and Channel 2 to point O (v_{out}). Then connect the signal generator to the circuit as shown in Figure 14–1, and adjust the sine wave output level of the generator at 0.2 V *peak-to-peak* at a frequency of 5 kHz.

   You should observe that the output signal level (v_{out}) is very nearly the same as the input level (v_{in}). In addition, there is no phase shift. These points are two major characteristics of a common-collector amplifier.
4. Using the *measured* value for the dc emitter voltage obtained in Step 2, calculate the dc emitter current I_E and the resultant transistor ac emitter resistance r_e (Equation 3). Record these values in Table 14–2.

5. Using the oscilloscope, measure the ac peak-to-peak voltage across the 1-kΩ load resistor. Calculate the voltage gain from base to emitter using Equation 2 given in the "Useful Formulas" section of this experiment. Record this result in Table 14–3. Now measure the actual voltage gain by dividing the peak-to-peak output voltage v_{out} by the peak-to-peak input voltage v_{in}, recording your result in Table 14–3.

6. Repeat Step 5 using the remaining load resistance values specified in Table 14–3. You should observe that the output voltage level decreases slightly. It does so because the load resistance affects the voltage gain of the amplifier stage. As in Step 5, experimentally determine the voltage gain by measuring v_{out} and v_{in}, comparing your measured result with the expected value (Equation 3). Record your results in Table 14–3.

WHAT YOU HAVE DONE

This experiment demonstrated the operation and characteristics of a small-signal common-collector amplifier, or *emitter–follower,* which has no phase shift. Here, the input signal is applied to the transistor's base lead, while the output signal is taken from the emitter lead. The experiment also showed how the load resistance influenced the circuit's voltage gain.

Name _____ Date _____

THE COMMON-COLLECTOR AMPLIFIER (EMITTER-FOLLOWER)

OBJECTIVES/PURPOSE:

SCHEMATIC DIAGRAM:

133

DATA FOR EXPERIMENT 14

TABLE 14–1

Parameter	Measured Value	Expected Value	% Error
V_B			
V_E			

TABLE 14–2

Parameter	Value
I_E (calculated)	
r_e (calculated)	

TABLE 14–3

Load Resistance	v_{in}	v_{out}	Measured Gain	Expected Gain	% Error
1 kΩ					
100 Ω					
68 Ω					

134

Name _____ Date _____

RESULTS AND CONCLUSIONS:

REVIEW QUESTIONS FOR EXPERIMENT 14

1. The voltage gain of a common-emitter amplifier, or an *emitter-follower*, is always
   (a) greater than 1   (b) equal to 1   (c) less than 1   (   )

2. The output signal of a common-collector amplifier is out-of-phase with the input by
   (a) 0°   (b) 45°   (c) 90°   (d) 180°   (   )

3. If the load resistor R_L in the circuit of Figure 14–1 is increased, the voltage gain will
   (a) increase significantly   (b) decrease significantly
   (c) remain essentially the same   (   )

4. For the voltage gain to approach 1,
   (a) R_L must be omitted   (b) R_L must be shorted
   (c) β must be as large as possible
   (d) r_e must be as small as possible   (   )

5. Which of the following is not characteristic of a common-collector amplifier?
   (a) 0° phase shift   (b) Voltage gain less than 1
   (c) Output taken from emitter   (d) Low input impedance   (   )

NOTES

15

THE COMBINATION COMMON-EMITTER AMPLIFIER AND EMITTER-FOLLOWER

PURPOSE AND BACKGROUND

The purpose of this experiment is to demonstrate the operation of a combination common-emitter amplifier and emitter-follower circuit. This type of circuit, sometimes referred to as a *phase-splitter* or a *paraphase amplifier,* produces two identical output signals to identical loads, except that they are 180° out-of-phase with each other. The output signal from the collector is simply a common-emitter amplifier whose voltage gain is 1 in addition to being 180° out-of-phase with the input signal. The output signal is from the emitter-follower and is in-phase with the input signal.

Text References: 6–2, Transistor AC Equivalent Circuits; 6–3, Common-Emitter Amplifiers.

REQUIRED PARTS AND EQUIPMENT

Resistors (1/4 W):
- ☐ Four 1 kΩ
- ☐ Two 10 kΩ

Capacitors (25 V):
- ☐ Two 2.2 μF
- ☐ 100 μF

- ☐ 2N3904 npn silicon transistor
- ☐ 0–15 V dc power supply
- ☐ Signal generator
- ☐ VOM or DMM (preferred)
- ☐ Dual trace oscilloscope
- ☐ Breadboarding socket

USEFUL FORMULAS

Voltage gain from base to collector

(1) $A_v = \dfrac{v_{\text{out1}}}{v_{\text{in}}}$

(2) $A_v = \dfrac{R_C \parallel R_{L1}}{(R_E \parallel R_{L2}) + r_e}$

Voltage gain from base to emitter

(3) $A_v = \dfrac{v_{\text{out2}}}{v_{\text{in}}}$

(4) $A_v = \dfrac{R_E \parallel R_{L2}}{(R_E \parallel R_{L2}) + r_e}$

Transistor ac emitter resistance (at normal room temperature)

(5) $r_e \cong \dfrac{25 \text{ mV}}{I_E}$

Quiescent dc base voltage

(6) $V_B = \left(\dfrac{R_2}{R_1 + R_2}\right) V_{CC}$

Quiescent dc emitter voltage

(7) $V_E = V_B - V_{BE}$

Quiescent dc emitter current

(8) $I_E = \dfrac{V_E}{R_E}$ ($I_C \simeq I_E$ for large β)

Quiescent dc collector voltage

(9) $V_C = V_{CC} - I_C R_C$

PROCEDURE

1. Wire the circuit shown in Figure 15–1, omitting the signal generator and the power supply.
2. After you have checked all connections, apply the 15-V supply voltage to the breadboard. With a VOM or DMM, individually measure the transistor dc base, collector, and emitter voltages with respect to ground, recording your results in Table 15–1. Based on the resistor values of Figure 15–1, determine the expected values of these three voltages (Equations 6, 7, and 9), assuming a base-emitter voltage drop of 0.7 V, and compare them with the measured values in Table 15–1.
3. Connect Channel 1 of your oscilloscope to point I (v_{in}) and Channel 2 to point OA (v_{out1}). Then connect the signal generator to

the circuit as shown in Figure 15–1, and adjust the sine wave output level of the generator at 0.5 V *peak-to-peak* at a frequency of 5 kHz.

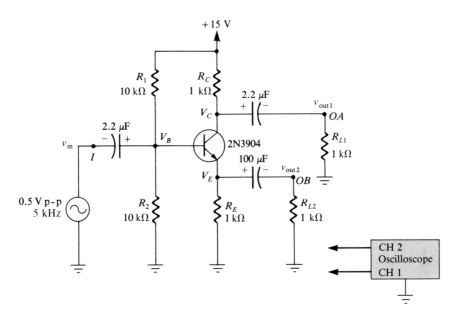

FIGURE 15–1 *Schematic diagram of circuit.*

You should observe that the output signal level (v_{out1}) is the same as the input level (v_{in}). In addition, there is a 180° phase shift between the output and input signals.

4. Using the *measured* value for the dc emitter voltage obtained in Step 2, calculate the dc emitter current (Equation 8) and the resultant ac emitter resistance, r_e (Equation 5). Record these values in Table 15-2.

5. Measure the ac peak-to-peak voltage across the 1-kΩ load resistor (R_{L1}) with the oscilloscope. Calculate the expected voltage gain from base to collector using Equation 2 in the "Useful Formulas" section of this experiment, and record this value in Table 15–3. Now measure the actual voltage gain by dividing the peak-to-peak output voltage v_{out1} by the peak-to-peak input voltage v_{in} (Equation 1), recording your result in Table 15–3.

6. Measure the ac peak-to-peak voltage across the 1-kΩ load resistor (R_{L2}) with the oscilloscope. Calculate the expected voltage gain from base to emitter using Equation 4 in the "Useful Formulas" section of this experiment, and record this value in Table 15–3. Now measure the actual voltage gain by dividing the peak-to-peak output voltage v_{out2} by the peak-to-peak input voltage v_{in} (Equation 3), recording your result in Table 15–3.

139

7. Connect Channel 1 of your oscilloscope to point OA (v_{out1}) and Channel 2 to point OB (v_{out2}). You should observe that both output signal levels are the same as the input level, except for a 180° phase shift between them.

WHAT YOU HAVE DONE

This experiment demonstrated the operation and characteristics of a combination common-emitter amplifier and emitter-follower circuit, which is sometimes referred to either as a *phase splitter* or a *paraphase amplifier*. This type of circuit produces two identical output signals to identical loads, except that they are 180° out-of-phase with each other.

Name ______________________________ Date ____________

THE COMBINATION COMMON-EMITTER AMPLIFIER AND EMITTER-FOLLOWER

OBJECTIVES/PURPOSE:

SCHEMATIC DIAGRAM:

Name _____   Date _____

DATA FOR EXPERIMENT 15

TABLE 15–1

Parameter	Measured Value	Expected Value	% Error
V_B			
V_E			
V_C			

TABLE 15–2

Parameter	Value
I_E (calculated)	
r_e (calculated)	

TABLE 15–3

Output	v_{in}	v_{out}	Phase Shift	Measured Gain	Expected Gain	% Error
OA						
OB						

Name _____ Date _____

RESULTS AND CONCLUSIONS:

REVIEW QUESTIONS FOR EXPERIMENT 15

1. The voltage gain at either output for the phase-splitter circuit of Figure 15-1 is
    (a) significantly less than 1   **(b)** essentially equal to 1
    (c) significantly greater than 1   (   )
2. The two output signals are out-of-phase with each other by
    (a) 0°   **(b)** 45°   **(c)** 90°   **(d)** 180°   (   )
3. If R_{L2} in the circuit of Figure 15–1 is omitted, v_{out1}
    (a) increases significantly   **(b)** decreases significantly
    (c) remains essentially the same   (   )
4. If R_{L1} in the circuit of Figure 15–1 is omitted, v_{out2}
    (a) increases significantly   **(b)** decreases significantly
    (c) remains essentially the same   (   )
5. If R_{L1} in the circuit of Figure 15–1 is omitted, v_{out1}
    (a) increases significantly   **(b)** decreases significantly
    (c) remains essentially the same   (   )

NOTES

16

THE COMMON-BASE AMPLIFIER

PURPOSE AND BACKGROUND

The purposes of this experiment are (1) to demonstrate the operation and characteristics of the common-base amplifier and (2) to investigate what influences its voltage gain. The common-base amplifier is characterized by application of the amplifier input signal to the emitter lead while its output is taken from the collector. Thus, as in the emitter-follower, both signals are in-phase. However, the voltage gain of a common-base amplifier is like that of a common-emitter amplifier.

Text Reference: 6–5, Common-Base Amplifiers.

REQUIRED PARTS AND EQUIPMENT

Resistors (1/4 W):
- [ ] 470 Ω
- [ ] Two 1 kΩ
- [ ] 10 kΩ

Capacitors (25 V):
- [ ] 2.2 μF
- [ ] 100 μF

- [ ] 2N3904 npn silicon transistor
- [ ] Two 0–15 V dc power supplies
- [ ] Signal generator
- [ ] VOM or DMM (preferred)
- [ ] Dual trace oscilloscope
- [ ] Breadboarding socket

USEFUL FORMULAS

Voltage gain from emitter to collector

(1) $A_v = \dfrac{v_{out}}{v_{in}}$

(2) $A_v = \dfrac{R_C \| R_L}{r_e}$ (normal circuit)

(3) $A_v = \dfrac{R_C}{r_e}$ (without load)

Transistor ac emitter resistance (at normal room temperature)

(4) $r_e \cong \dfrac{25 \text{ mV}}{I_E}$

Quiescent dc emitter voltage

(5) $V_E = -V_{BE}$

Quiescent dc emitter current

(6) $I_E = \dfrac{V_{EE} - V_{BE}}{R_E}$ ($I_E \simeq I_C$ for large β)

Quiescent dc collector voltage

(7) $V_C = V_{CC} - I_C R_C$

PROCEDURE

1. Wire the circuit shown in Figure 16–1, omitting the signal generator and the power supply.
2. After you have checked all connections, apply the +9-V and −9-V supply voltages to the breadboard. With either a VOM or DMM, individually measure the transistor dc emitter and collector voltages with respect to ground, recording your results in Table 16–1. Based on the resistor values of Figure 16–1, determine the expected values of these two voltages (Equations 5 and 7), assuming a base-emitter voltage drop of 0.7 V, and compare them with the measured values in Table 16–1.
3. Connect Channel 1 of your oscilloscope to point I (v_{in}) and Channel 2 to point O (v_{out}). Then connect the signal generator to the circuit as shown in Figure 16–1, and adjust the sine wave output level of the generator at 25 mV *peak-to-peak* at a frequency of 5 kHz. If you cannot reach 25 mV, then adjust v_{in} so that there is no clipping on the output signal.

   You should observe that the output signal level (v_{out}) is *greater* than the input level (v_{in}). In addition, v_{out} is in-phase with respect to the input. These points are two major characteristics of a common-base amplifier.

4. Using the measured value for the dc emitter voltage obtained in Step 2, calculate the dc quiescent emitter current (Equation 6) and the resultant transistor ac emitter resistance, r_e (Equation 4). Record these values in Table 16–2.

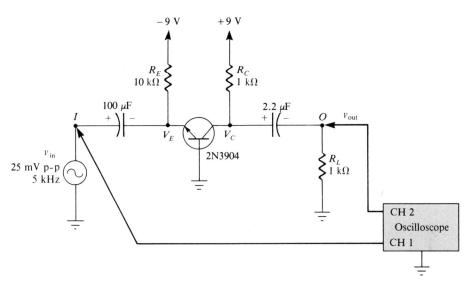

FIGURE 16–1 *Schematic diagram of circuit.*

5. Calculate the voltage gain from emitter to collector using Equation 2 given in the "Useful Formulas" section of this experiment, and record the value in Table 16–3. Now measure the actual voltage gain by dividing the peak-to-peak output voltage v_{out} by the peak-to-peak input voltage v_{in} (Equation 1), recording your result in Table 16–3.
6. Remove R_L. You should observe that the output voltage level increases. It does so because the load resistance affects the voltage gain of the amplifier stage. As in Step 5, experimentally determine the voltage gain by measuring v_{out} and v_{in}, comparing your measured result with the expected value (Equation 3). Record your results in Table 16–3.
7. Finally, connect a 470-Ω resistor for R_L. Calculate the voltage gain from emitter to collector using Equation 2 given in the "Useful Formulas" section of this experiment, and record the value in Table 16–3. Now measure the actual voltage gain by dividing the peak-to-peak output voltage v_{out} by the peak-to-peak input voltage v_{in}, recording your result in Table 16–3.

From the results in Table 16–3, you should now understand how the load resistance affects the emitter-to-collector voltage gain of a common-base amplifier.

147

WHAT YOU HAVE DONE

This experiment demonstrated the operation and characteristics of a small-signal common-base amplifier, which has no phase shift. Here, the input signal is applied to the transistor's emitter lead, while the output signal is taken from the collector lead. The experiment also showed how the load resistance influenced the circuit's voltage gain.

Name _____   Date _____

THE COMMON-BASE AMPLIFIER

OBJECTIVES/PURPOSE:

SCHEMATIC DIAGRAM:

149

DATA FOR EXPERIMENT 16

TABLE 16–1

Parameter	Measured Value	Expected Value	% Error
V_E			
V_C			
I_E			
r_e			

TABLE 16–2

Parameter	Value
I_E (calculated)	
r_e (calculated)	

TABLE 16–3

Load Resistance	v_{in}	v_{out}	Measured Gain	Expected Gain	% Error
1 kΩ					
None					
470 Ω					

RESULTS AND CONCLUSIONS:

REVIEW QUESTIONS FOR EXPERIMENT 16

1. For the circuit of Figure 16–1, the voltage gain from base to
   collector is approximately
   (a) 1   (b) 17   (c) 34   (d) 50   (   )
2. The output signal is out-of-phase with the input by
   (a) 0°   (b) 45°   (c) 90°   (d) 180°   (   )
3. If the emitter supply voltage is made more negative, the voltage
   gain will
   (a) increase
   (b) decrease
   (c) remain essentially the same   (   )
4. For the circuit of Figure 16–1, V_E is approximately
   (a) −9 V   (b) 0 V   (c) −0.7 V   (d) +0.7 V   (   )
5. Which of the following is not a normal characteristic of a common-
   base amplifier:
   (a) 180° phase shift
   (b) low input impedance
   (c) output taken from collector
   (d) voltage gain greater than 1   (   )

NOTES

17

THE CLASS A COMMON-EMITTER POWER AMPLIFIER

PURPOSE AND BACKGROUND

The purpose of this experiment is to demonstrate the operation of a Class A common-emitter power amplifier. The Class A amplifier is biased such that collector current always flows during the entire cycle of the input waveform. Ideally, the amplifier's Q point should be biased at the center of the ac load line so that the output signal can have the maximum possible swing in both directions. Consequently, clipping will occur simultaneously on both peaks of the output signal if the amplifier is overdriven. If the Q point is not centered on the ac load line, output waveform clipping will occur first, either at saturation or at cutoff. Despite the simplicity of the Class A amplifier, the maximum efficiency that can be expected for it with a capacitively coupled load is only 25 percent.

Text Reference: 7–1, Class A amplifiers.

REQUIRED PARTS AND EQUIPMENT

Resistors (1/4 W):
- ☐ 220 Ω
- ☐ 560 Ω
- ☐ 1 kΩ
- ☐ 4.7 kΩ
- ☐ 10 kΩ
- ☐ 100 kΩ
- ☐ 5-kΩ potentiometer, or 10-turn "trimpot"

Capacitors (25 V):
- ☐ Two 2.2 μF
- ☐ 100 μF
- ☐ 2N3904 npn silicon transistor
- ☐ 0–15 V dc power supply
- ☐ VOM or DMM (preferred)
- ☐ Signal generator
- ☐ Dual trace oscilloscope
- ☐ Breadboarding socket

USEFUL FORMULAS

Quiescent dc base voltage

$$(1) \quad V_B = \left(\frac{R_2}{R_1 + R_2} \right) V_{CC}$$

Quiescent dc emitter voltage

$$(2) \quad V_E = V_B - V_{BE}$$

Quiescent collector current

$$(3) \quad I_{CQ} = \frac{V_E}{R_E} \qquad (I_E \simeq I_{CQ} \text{ for large } \beta)$$

Quiescent dc collector-emitter voltage

$$(4) \quad V_{CEQ} = V_{CC} - I_{CQ}(R_C + R_E)$$

ac load line

Collector saturation current:

$$(5) \quad I_{c(\text{sat})} = I_{CQ} + \frac{V_{CEQ}}{R_c}$$

Collector-emitter cutoff voltage:

$$(6) \quad V_{ce(\text{cutoff})} = V_{CEQ} + I_{CQ}R_c$$

For centered Q point

$$(7) \quad I_{c(\text{sat})} = 2I_{CQ}$$

$$(8) \quad V_{CE(\text{cutoff})} = 2V_{CEQ}$$

$$(9) \quad R_c \cong \frac{V_{CEQ}}{I_{CQ}}$$

where $R_c = R_c \| R_L$

rms output (load) power

$$(10) \quad P_o(\text{rms}) = \frac{[V_o(\text{rms})]^2}{R_L}$$

dc power supplied to amplifier

(11) $P_{dc} = V_{CC}I_{CQ}$

Amplifier percent efficiency

(12) $\%\eta = \dfrac{P_o(\text{rms})}{P_{dc}} \times 100\%$

PROCEDURE

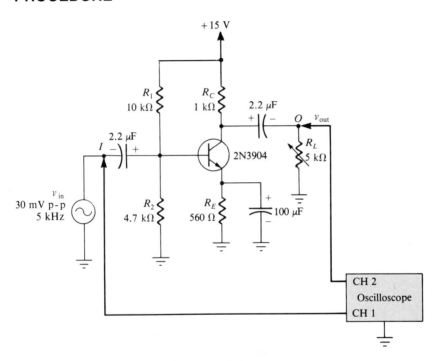

FIGURE 17–1 *Schematic diagram of circuit.*

1. Wire the circuit shown in Figure 17–1, omitting the signal generator and the power supply.
2. After you have checked all connections, apply only the 15-V supply voltage to the breadboard. With the VOM or DMM, individually measure the transistor dc base and emitter voltages with respect to ground, as well as V_{CEQ}, recording your results in Table 17–1. Based on the resistor values of Figure 17–1, determine the expected values of these three voltages, assuming a base-emitter voltage drop of 0.7 V, and compare them with the measured values in Table 17–1.
3. Measure the quiescent dc collector current I_{CQ}, comparing it with the expected value (Equation 3). Record your values in Table 17–1.

155

4. Connect Channel 1 of your oscilloscope at point I (v_{in}) and Channel 2 to point O (v_{out}). Adjust your oscilloscope to the following approximate settings:

    Channel 1: 10 mV/division, ac coupling
    Channel 2: 1 V/division, ac coupling
    Time base: 0.2 ms/division

5. From Equation 9 given in the "Useful Formulas" section of this experiment, determine the value of the load resistor R_L that gives a centered Q point on the ac load line. Adjust the 5-kΩ potentiometer load as closely as you can to that value.

6. Connect the signal generator to the breadboard, and adjust the sine wave output level of the generator at 30 mV *peak-to-peak* at a frequency of 5 kHz. If your signal generator cannot be adjusted to 30 mV peak-to-peak, any value higher than this can be used as long as the output signal is not clipped. You should observe that the peak-to-peak output voltage is much larger than the input, in addition to having a phase shift of 180°. Slowly increase the output level of the signal generator. What eventually happens to the output waveform?

    After a point, the peak-to-peak output voltage no longer increases. In fact, the positive and negative peaks become flattened, or *clipped*. Note that if the Q point is placed very near the center of the ac load line, both peaks become clipped at approximately the same time. Consequently, the transistor reaches cutoff and saturation at the same input level.

7. Now reduce the input signal level to zero, and replace the potentiometer with a 220-Ω resistor. Slowly increase the input signal level so that one peak slips off much earlier than the other, as shown in Figure 17–2.

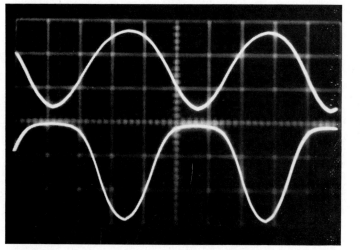

Input

Output

FIGURE 17–2

You should observe that only the *positive* peaks are clipped. This condition is characteristic of *cutoff clipping* because the Q point is closer to the ac load line's collector-emitter cutoff voltage than to the collector saturation current. Confirm this situation by drawing the ac load line with the Q point on the graph page provided.

8. Now reduce the input signal level to zero, and replace the 220-Ω resistor with a 100-kΩ resistor. Slowly increase the input signal level so that one peak clips off much earlier than the other, as shown in Figure 17–3.

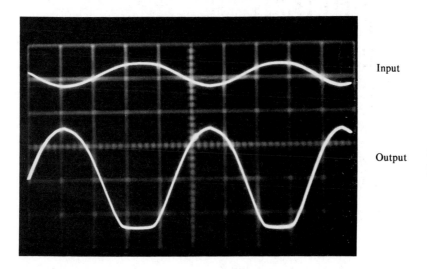

Input

Output

FIGURE 17–3

You should observe that now only the *negative* peaks are clipped. This condition is characteristic of *saturation clipping* because the Q point is closer to the ac load line's collector saturation current than to the collector-emitter's cutoff voltage. Confirm this situation by drawing the ac load line with the Q point on the graph page provided. You should have a different load line than that of Step 7.

9. As in Step 5, again adjust the 5-kΩ potentiometer to the resistance that centers the Q point on the ac load line, replacing the 100-kΩ resistor with it.

10. Now carefully increase the peak-to-peak input signal just before both output peaks clip off. With your VOM or DMM measure the rms voltage across the load resistor, V_o (rms), and compute the rms output power of the amplifier (Equation 10). Record these results in Table 17–2.

11. Using Equation 11 given in the "Useful Formulas" section of the experiment, calculate the dc power supplied, recording this value in Table 17–2.

12. Finally, compute the percent efficiency (%η) of your amplifier (Equation 12), and compare it with the theoretical maximum of 25%. Record your result in Table 17–2. If you calculate a value greater than 25%, then repeat Steps 9, 10, and 11, trying to determine the source of your error.

WHAT YOU HAVE DONE

This experiment demonstrated the operation and characteristics of Class A common-emitter power amplifier. Here, it was shown how the location of the amplifier Q point affects the type of clipping observed on the output signal when the amplifier circuit is overdriven. If the Q point was closer to saturation, then the negative peak was clipped first. The opposite was found to be true if biased closer to cutoff. In addition, the efficiency of the amplifier was determined.

Name ______________________ Date ____________

THE CLASS A COMMON-EMITTER POWER AMPLIFIER

OBJECTIVES/PURPOSE:

SCHEMATIC DIAGRAM:

Name _____   Date _____

DATA FOR EXPERIMENT 17

TABLE 17–1 *Class A amplifier bias parameters.*

Parameter	Measured Value	Expected Value	% Error
V_B			
V_E			
V_{CEQ}			
I_{CQ}			

TABLE 17–2 *Class A amplifier efficiency.*

Parameter	Measured Value
V_o (rms)	
Parameter	Calculated Value
P_o (rms)	
P_{dc}	
$\%\eta$	

Name _____ Date _____

DATA FOR EXPERIMENT 17

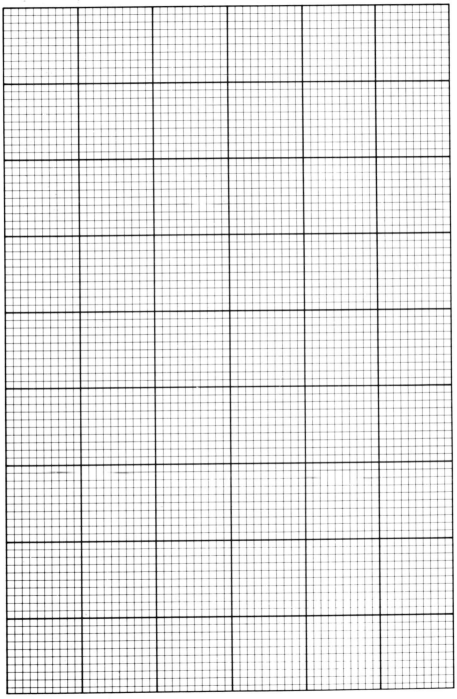

Name _____ Date _____

RESULTS AND CONCLUSIONS:

REVIEW QUESTIONS FOR EXPERIMENT 17

1. For the circuit of Figure 17–1, the value of R_L that centers the
   Q point is approximately
   (a) 500 Ω   **(b)** 750 Ω   **(c)** 1 kΩ   **(d)** 10 kΩ   (   )
2. For the best performance, the Q point should be centered on
   (a) the dc load line   **(b)** the ac load line
   (c) both the ac and dc load lines   (   )
3. If the negative peaks of the output waveform are clipped as
   shown in Figure 17–3, this condition represents
   (a) cutoff clipping   **(b)** saturation clipping
   (c) nonlinear distortion   (   )
4. Making R_L much larger than 1 kΩ would result in
   (a) cutoff clipping of the output signal
   (b) saturation clipping of the output signal
   (c) nonlinear distortion of the output signal   (   )
5. For the Class A amplifier of Figure 17–1, the maximum peak-to-
   peak output voltage that can be obtained without clipping with
   a 500-Ω load is approximately
   (a) 5 V   **(b)** 7.5 V   **(c)** 10 V   **(d)** 15 V   (   )

162

18

THE CLASS B PUSH-PULL EMITTER-FOLLOWER POWER AMPLIFIER

PURPOSE AND BACKGROUND

The purpose of this experiment is to demonstrate the design and operation of a Class B push-pull emitter-follower power amplifier. The Class B push-pull amplifier has a pair of complementary (npn and pnp) transistors, each of which is biased at cutoff (that is, no collector current). Consequently, collector current in each transistor flows only for alternate half-cycles of the input signal. Since both transistors are biased at cutoff, the input signal must be sufficient to forward bias each transistor on the appropriate half-cycle of the input waveform. As a result, crossover distortion occurs.

To eliminate crossover distortion, both transistors should, under quiescent conditions, be slightly forward biased so that each transistor is actually biased slightly before cutoff, resulting in a small amount of current called the *trickle current*. Since we now have neither true Class A nor true Class B operation, but rather something in between, this operation is referred to as *Class AB operation*, although the term "Class B" is frequently used to describe this situation. Despite its complexity, a Class B push-pull amplifier can achieve efficiencies up to approximately 78%, which is more than three times better than can be obtained with a similar Class A amplifier without transformer coupling.

Text Reference: 7–2, Class B and AB Push-Pull Amplifiers.

REQUIRED PARTS AND EQUIPMENT

Resistors (1/4 W):
- [ ] 100 Ω
- [ ] Three 1 kΩ
- [ ] Two 10 kΩ
- [ ] 5-kΩ potentiometer, or 10-turn "trimpot"

Capacitors (25 V):
- [ ] Two 2.2 μF
- [ ] 100 μF

- [ ] Two 1N914 (or 1N4148) silicon diodes
- [ ] 2N3904 npn silicon transistor
- [ ] 2N3906 pnp silicon transistor
- [ ] 0–15 V dc power supply
- [ ] VOM or DMM (preferred)
- [ ] Signal generator
- [ ] Dual trace oscilloscope
- [ ] Breadboarding socket

USEFUL FORMULAS

Quiescent dc collector current (diode bias)

$$(1) \quad I_{CQ} = \frac{V_{CC} - 2V_{BE}}{R_1 + R_4} \qquad (\text{if } V_{BE1} = V_{BE2})$$

rms output power

$$(2) \quad P_o(\text{rms}) = \frac{[V_o(\text{rms})]^2}{R_L}$$

dc supply power supplied to amplifier

$$(3) \quad P_{dc} = V_{CC}I_{CC}$$

Amplifier percent efficiency

$$(4) \quad \%\eta = \frac{P_o(\text{rms})}{P_{dc}} \times 100\%$$

PROCEDURE

1. Wire the circuit shown in Figure 18–1A. Connect Channel 1 of your oscilloscope at point $I(v_{in})$ and Channel 2 to point $O(v_{out})$. Adjust your oscilloscope to the following approximate settings:

   Channels 1 and 2: 1 V/division, ac coupling
   Time base: 0.2 ms/division

2. Apply power to the breadboard and adjust the sine wave output level of the generator at 6 V *peak-to-peak* at a frequency of 1 kHz. You should observe amplifier input and output waveforms similar to those shown in Figure 18–2. Notice that the output waveform is distorted in the vicinity of zero volts. This condition, referred to as *crossover distortion*, results when the base-emitter diodes of both transistors are not forward biased

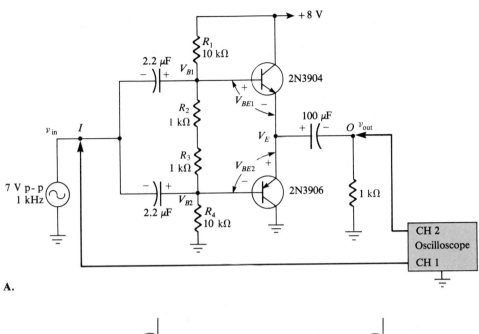

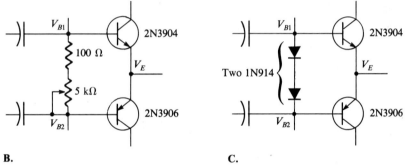

FIGURE 18–1 *Schematic diagram of circuits.*

until the input signal exceeds approximately 0.7 V in both directions.

Using Figure 18–3 as a guide, measure the base-to-emitter voltages required for both transistors to become forward biased, recording these values in Table 18–1.

Note that the peak-to-peak output voltage is slightly smaller than the input, a difference approximately equal to two base-emitter voltage drops. In addition, since each half of the push-pull circuit is itself an emitter-follower, there is no phase shift between the input and output signals.

3. Disconnect the power and signal generator leads from the breadboard, and replace the series resistance $R_2 + R_3$ with a 5-kΩ potentiometer in series with 100-Ω resistor, as shown in Figure 18–1B. Again connect the power and signal generator leads to the breadboard.

165

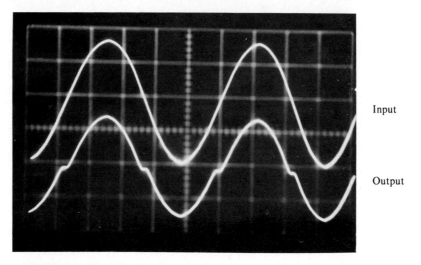

Input

Output

FIGURE 18–2

4. With the signal generator set at 6 V peak-to-peak at 1 kHz, carefully adjust the potentiometer so that the crossover distortion just disappears. Notice that the output voltage very nearly equals the input, so the voltage gain is essentially equal to 1.

5. Again disconnect the power and signal generator leads from the breadboard, and replace the potentiometer and 100-Ω resistor with two 1N914 (or 1N4148) compensating diodes in series between the two transistor base leads, as shown in Figure 18–1C. Again connect the power and signal generator to the breadboard.

6. With the signal generator set at 6 V peak-to-peak at 1 kHz, how does the output signal compare with that seen in Step 4, after the potentiometer is adjusted?

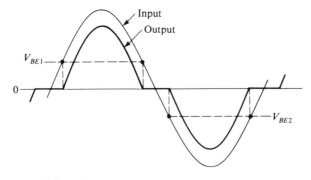

FIGURE 18–3

You should find that there is virtually no crossover distortion. The voltage required to forward bias both transistors is now supplied by the voltage drops of the two silicon diodes, which are also forward biased.

7. With a VOM or DMM, individually measure the transistor dc base 1, base 2, and emitter voltages with respect to ground, recording your results in Table 18–2.

8. Now carefully increase the peak-to-peak input signal so that the output peaks just clip off. With your VOM or DMM, measure the rms voltage across the 1-kΩ load resistor [V_o(rms)], and compute the rms output power of the amplifier (Equation 2). Record these results in Table 18–3.

9. In order to measure the dc power supplied to the amplifier while amplifying an input signal, use your VOM or DMM to measure the *dc* collector current I_{CC} of either transistor. Compute the dc power supplied (Equation 3), recording your results in Table 18–3.

10. Finally, compute the percent efficiency (%η) of your amplifier, and compare it with the theoretical maximum of 78.5% of a Class B amplifier. Record your result in Table 18–3. If you calculate a value greater than 78.5%, then repeat steps 8 and 9, trying to determine the source of your error.

WHAT YOU HAVE DONE

This experiment demonstrated the design and operation of a Class B push-pull emitter-follower power amplifier and how it compared with that of a Class A amplifier circuit. The circuit required a complementary pair of transistors (one each pnp and npn) have almost identical characteristics. The causes of crossover distortion were demonstrated and possible methods to eliminate it were investigated. The most effective of these methods was current-mirror biasing using a pair of diodes. In addition, the efficiency of the amplifier was determined and compared with that of a Class A amplifier.

NOTES

Name ______________________________ Date ____________

THE CLASS B PUSH-PULL EMITTER-FOLLOWER POWER AMPLIFIER

OBJECTIVES/PURPOSE:

SCHEMATIC DIAGRAM:

DATA FOR EXPERIMENT 18

TABLE 18–1 *Voltage-divider bias with no crossover distortion.*

Parameter	Measured Value
V_{BE1}	
V_{BE2}	
V_E	

TABLE 18–2 *Diode (current mirror) bias with no crossover distortion.*

Parameter	Measured Value
V_{B1}	
V_{B2}	
V_E	

TABLE 18–3 *Class B amplifier efficiency.*

Parameter	Measured Value
V_o (rms)	
I_{CC}	
Parameter	Calculated Value
P_o (rms)	
P_{dc}	
$\%\eta$	

Name _____   Date _____

RESULTS AND CONCLUSIONS:

REVIEW QUESTIONS FOR EXPERIMENT 18

1. The output distortion shown in Figure 18–2 is called
   (a) *harmonic* distortion (b) *nonlinear* distortion
   (c) *crossover* distortion (d) *amplitude* distortion   (   )
2. For the Class B amplifier circuit of Figure 18–1, crossover distortion generally occurs when
   (a) both transistors are biased at cutoff
   (b) the base-emitter junctions of both transistors are not forward biased
   (c) both (a) and (b)   (   )
3. Elimination of crossover distortion for the circuit of Figure 18–1 can be achieved by
   (a) adding compensation diodes that forward bias the base-emitter junction of both transistors
   (b) increasing the supply voltage
   (c) adjusting resistors R_1, R_2, R_3, and R_4 to forward bias the base-emitter junction of both transistors
   (d) all of the above   (   )
4. If crossover distortion is eliminated, then the amplifier operation of Figure 18–1 is more correctly termed
   (a) Class A (b) Class AB (c) Class B (d) Class C   (   )
5. The efficiency of a Class B amplifier cannot be greater than approximately
   (a) 25% (b) 50% (c) 75% (d) 100%   (   )

171

NOTES

19

THE JFET DRAIN CURVE

PURPOSE AND BACKGROUND

The purpose of this experiment is to use the oscilloscope to display the drain curve for the MPF102 JFET. This curve shows the variation of drain current as a function of drain-to-source voltage with a constant gate-to-source voltage. Thus, with the special case of having zero gate-to-source voltage, it is possible to estimate JFET parameters such as drain current with gate shorted to source, I_{DSS}, as well as gate-to-source cutoff voltage, $V_{GS(off)}$.

   Text Reference: 8–2, JFET Characteristics and Parameters.

REQUIRED PARTS AND EQUIPMENT

☐ 100-Ω resistor, 1/4 W
☐ MPF102 n-channel JFET
☐ 1N4001 silicon rectifier diode
☐ 12.6-V rms secondary transformer
☐ Dual trace oscilloscope
☐ Breadboarding socket

PROCEDURE

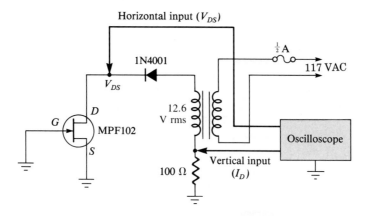

FIGURE 19–1 *Schematic diagram of circuit.*

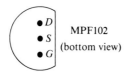

FIGURE 19–2 *Pin configuration of MPF102 JFET.*

1. Wire the circuit shown in the schematic diagram of Figure 19–1. In this part, the oscilloscope is set up to function as an X-Y plotter. Set the oscilloscope's controls to the following approximate settings:

    Vertical (or Y) input sensitivity: 0.2 V/division, dc coupling

    Horizontal (or X) input sensitivity: 1 V/division, dc coupling

2. After the oscilloscope has warmed up, move the trace dot to the upper left-hand corner of the oscilloscope's screen so that it is centered on the intersecting scale divisions of the display. Now connect the transformer's primary to the 117-V ac power line. The oscilloscope display should be similar to that shown in Figure 19–3. If it is not, the oscilloscope leads may be interchanged or there may be a wiring error.

    The "horizontal input" measures the JFET's instantaneous drain-to-source voltage (V_{DS}). The diode allows only positive

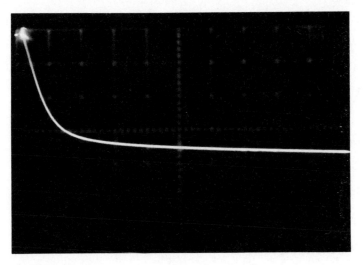

FIGURE 19–3

voltage variations, from 0 V to about 18 V, which serves as the instantaneous drain-to-source voltage. The "vertical input" measures the voltage drop across the 100-Ω resistor. As shown on the display, the vertical axis increases *downward,* which is displayed inverted from the normal sense. Using Ohm's law, we can make the vertical input read the JFET's instantaneous drain current (I_D). If, for example, the vertical sensitivity is 0.2 V/division, then, in terms of the current through the 100-Ω resistor (which is the same as the drain current),

$$\text{Vertical sensitivity} = \frac{0.2 \text{ V/division}}{100 \ \Omega}$$

$$= 2 \text{ mA/division}$$

3. As shown in the schematic diagram of Figure 19–1, both the gate and source leads are grounded, and thus V_{GS} is zero. Consequently, the maximum drain current (I_{DSS}) occurs when the curve flattens out. The pinch-off region of the drain curve is for drain-to-source voltages greater than the JFET's *pinch-off* voltage, V_p. This also occurs when the drain current just begins to flatten out. Numerically, the pinch-off voltage is equal to the JFET's gate-to-source cutoff voltage, $V_{GS(off)}$. Using Figure 19–4 as a guide, estimate from the oscilloscope's display both I_{DSS} and $V_{GS(off)}$ for the MPF102 JFET you are using. Record these values in Table 19–1.

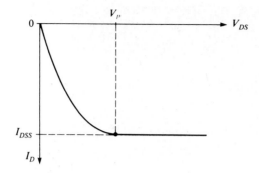

FIGURE 19–4

WHAT YOU HAVE DONE

This experiment demonstrated the shape of the drain curve of a MPF102 JFET whose gate-to-source voltage is zero. Using an oscilloscope as a simple curve tracer, the drain characteristic shows the variation of the JFET's drain current as a function of drain-to-source voltage. This curve then allowed the determination of JFET parameters such as I_{DSS} and $V_{GS(off)}$.

Name ______________________ Date __________

THE JFET DRAIN CURVE

OBJECTIVES/PURPOSE:

SCHEMATIC DIAGRAM:

Name _____   Date _____

DATA FOR EXPERIMENT 19

TABLE 19–1 *MPF102 drain curve.*

$V_{GS} = 0$	
I_{DSS}	mA
$V_{GS(\text{off})}$	V

Name _____   Date _____

RESULTS AND CONCLUSIONS:

REVIEW QUESTIONS FOR EXPERIMENT 19

1. For the circuit of Figure 19–1, if the vertical sensitivity of the oscilloscope is set at 50 mV/division, the vertical axis is then calibrated to read
    (a) 0.5 mA/division   (b) 5 mA/division
    (c) 50 mA/division   (d) 0.5 A/division   (   )

2. The curve of Figure 19–1 is called a
    (a) JFET forward transconductance curve
    (b) JFET drain curve
    (c) JFET source curve
    (d) JFET gate curve   (   )

3. The point at which the curve of Figure 19–3 starts to flatten out determines
    (a) I_{DSS} (b) V_p (c) $V_{GS(off)}$ (d) all of the above   (   )

4. Which of the following does not apply to the curve of Figure 19–3?
    (a) valid for V_{GS} greater than 0 V   (b) can determine V_p
    (c) can determine $V_{GS(off)}$ (d) can determine I_{DSS} (   )

5. The JFET pinch-off voltage can be taken to be equal to
    (a) V_{GS} (b) $-V_{GS}$ (c) $V_{GS(off)}$ (d) $-V_{GS(off)}$ (   )

NOTES

20

THE JFET TRANSFER CHARACTERISTIC CURVE

PURPOSE AND BACKGROUND

The purpose of this experiment is to use the oscilloscope to display the transfer characteristic curve for the MPF102 JFET. This curve shows the parabolic, or square-law, variation of the drain current as a function of the gate-to-source voltage. From such a curve it is possible to estimate JFET parameters such as the drain current with gate shorted to source, I_{DSS}; the gate-to-source cutoff voltage, $V_{GS(off)}$; and the forward transconductance, g_{m0}. The values of these parameters are used in performing Experiments 21, 25, and 26.

Text Reference: 8–2, JFET Characteristics and Parameters.

REQUIRED PARTS AND EQUIPMENT

- ☐ 100-Ω resistor, 1/4 W
- ☐ MPF102 n-channel JFET
- ☐ 1N914 silicon rectifier diode
- ☐ 0–15 V dc power supply
- ☐ Signal generator
- ☐ Dual trace oscilloscope
- ☐ Breadboarding socket

USEFUL FORMULA

JFET forward transconductance at $V_{GS} = 0$

$$g_{m0} = \frac{2I_{DSS}}{|V_{GS(\text{off})}|}$$

PROCEDURE

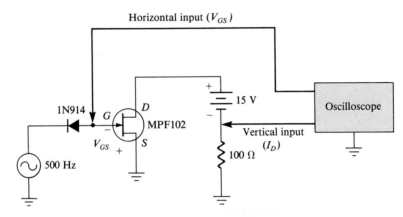

FIGURE 20–1 *Schematic diagram of circuit.*

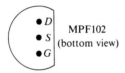

FIGURE 20–2 *Pin configuration of MPF102 JFET.*

1. Wire the circuit shown in the schematic diagram of Figure 20–1. In this part, the oscilloscope is set up to function as an X-Y plotter. Set the oscilloscope's controls to the following approximate settings:

   Vertical (or Y) input sensitivity: 0.2 V/division,
   dc coupling

   Horizontal (or X) input sensitivity: 1 V/division,
   dc coupling

2. After the oscilloscope has warmed up, move the trace dot to the upper right-hand corner of the oscilloscope's screen so that it is centered on the intersecting scale divisions of the display. Now apply power and the signal generator to the breadboard.

3. Adjust the frequency of the signal generator to 500 Hz and at a signal level sufficient to produce a display similar to that shown in Figure 20–3. If you do not obtain this curve, the oscilloscope leads may be interchanged or there may be wiring error.

The "horizontal input" measures the JFET's instantaneous gate-to-source voltage (V_{GS}). The diode allows only negative voltage variations, which serve as the drain-to-source voltage. The "vertical input" measures the voltage drop across the 100-Ω resistor. As shown on the display, the vertical axis increases *downward*, which is inverted from the normal sense. Using Ohm's law, we can make the vertical input read the JFET's instantaneous drain current (I_{DSS}). If the vertical sensitivity is 0.2 V/division, then, in terms of the current through the 100-Ω resistors (which is the same as the drain current),

$$\text{Vertical sensitivity} = \frac{0.2 \text{ V/division}}{100 \text{ } \Omega}$$

$$= 2 \text{ mA/division}$$

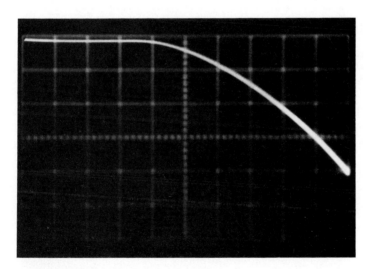

FIGURE 20–3

4. Using Figure 20–4 as a guide, estimate from the oscilloscope's display both I_{DSS} and $V_{GS(off)}$ for the MPF102 JFET you are using. Record these values in Table 20–1. If you also performed Experiment 19, now compare these values with those obtained in that experiment. If you used the same MPF102 JFET in both experiments, your results should closely agree.

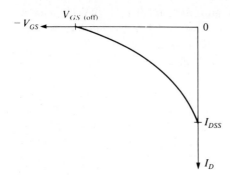

FIGURE 20–4

5. From the values for I_{DSS} and $V_{GS(off)}$ measured in Step 4, calculate the MPF102 JFET forward transconductance at zero gate-to-source voltage, g_{m0}, and record this value in Table 20–1. These values are used in performing Experiments 21, 25, and 26.

WHAT YOU HAVE DONE

This experiment demonstrated the shape of the transfer characteristic, or *transconductance* curve of an MPF102 JFET. Using an oscilloscope as a simple curve tracer, the transfer characteristic curve resembled a parabola showing the nonlinear variation of the JFET's drain current as a function of gate-to-source voltage. From this curve, it was then possible to estimate JFET parameters such as g_{m0}, I_{DSS}, and $V_{GS(off)}$, which would be needed for several later experiments.

Name ______________________ Date ____________

THE JFET TRANSFER CHARACTERISTIC CURVE

OBJECTIVES/PURPOSE:

SCHEMATIC DIAGRAM:

Name _____   Date _____

DATA FOR EXPERIMENT 20

TABLE 20–1 *MPF102 transfer characteristic curve.*

I_{DSS}	mA
$V_{GS(\text{off})}$	V
g_{m0}	μS

Name _____ Date _____

RESULTS AND CONCLUSIONS:

REVIEW QUESTIONS FOR EXPERIMENT 20

1. The point at which the curve of Figure 20–3 intersects the vertical axis is
    (a) g_{m0} **(b)** V_p **(c)** I_{DSS} **(d)** $V_{GS(off)}$ (   )
2. The point at which the curve of Figure 20–3 intersects the horizontal axis is
    (a) g_{m0} **(b)** V_{DS} **(c)** I_{DSS} **(d)** $V_{GS(off)}$ (   )
3. The diode in the test circuit of Figure 20–1 is used to
    (a) limit the peak drain current
    (b) forward bias the gate-to-source junction
    (c) reverse bias the gate-to-source junction
    (d) protect the JFET from excessive gate voltages   (   )
4. From the curve of Figure 20–3, one is able to determine
    (a) I_{DSS} **(b)** $V_{GS(off)}$ **(c)** g_{m0} **(d)** all of the above   (   )
5. If $g_{m0} = 5000\ \mu S$ and $V_{GS(off)} = -3$ V, then I_{DSS} is
    (a) 0.6 mA   **(b)** 1.2 mA   **(c)** 7.5 mA   **(d)** 15 mA   (   )

        187

NOTES

21

JFET SELF-BIAS

PURPOSE AND BACKGROUND

The purpose of this experiment is to verify the voltages and currents in a JFET circuit using self-bias. A JFET requires that the gate-to-source voltage always be less negative than the pinch-off or gate-to-source cutoff voltage, but less than zero. Since virtually no gate current flows due to the JFET's high input impedance, the gate voltage is essentially at ground reference. Consequently, using only a drain-supply voltage, the required negative quiescent gate-to-source voltage is developed by the voltage drop across the source resistor of the self-bias circuit. This experiment uses the JFET parameters measured in Experiment 20 to confirm measured values.

Text Reference: 8–3, JFET Biasing.

REQUIRED PARTS AND EQUIPMENT

Resistors (1/4 W):
- ☐ Two 1 kΩ
- ☐ 100 kΩ
- ☐ MPF102 n-channel JFET

- ☐ 0–15 V dc power supply
- ☐ VOM or DMM (preferred)
- ☐ Breadboarding socket

USEFUL FORMULAS

JFET dc gate-to-source cutoff voltage

$$(1) \quad V_{GS(\text{off})} = -\frac{2I_{DSS}}{g_{m0}}$$

Quiescent dc drain (source) current

$$(2a) \quad I_D = 2I_{DSS}\left[\frac{(R_S g_{m0} + 1) - (2R_S g_{m0} + 1)^{1/2}}{(R_S g_{m0})^2}\right]$$

$$(2b) \qquad = I_{DSS}\left[1 - \frac{V_{GS}}{V_{GS(\text{off})}}\right]^2$$

Quiescent dc gate-to-source voltage

$$(3) \quad V_S \quad = I_D R_S$$

$$= -V_{GS} \qquad \text{so that } V_G \simeq 0$$

Quiescent drain voltage

$$(4) \quad V_D = V_{DD} - I_D R_D$$

Quiescent dc drain-to-source voltage

$$(5) \quad V_{DS} = V_{DD} - I_D(R_D + R_S)$$

JFET forward transconductance at **Q** *point*

$$(6) \quad g_m = g_{m0}\left[1 - \frac{V_{GS}}{V_{GS(\text{off})}}\right]$$

PROCEDURE

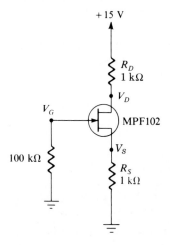

+ 15 V

R_D
1 kΩ

V_D

V_G

MPF102

100 kΩ

V_S

R_S
1 kΩ

FIGURE 21–1 *Schematic diagram of circuit.*

1. Wire the circuit shown in the schematic diagram of Figure 21–1, and apply power to the breadboard. For the same JFET used in Experiment 20, record the measured values for I_{DSS}, $V_{GS(off)}$, and G_{m0} in Table 21–1.
2. Using your VOM or DMM as an ammeter, measure the quiescent drain current (I_D) and record this value in Table 21–2.
3. Using the values of g_{m0} and I_{DSS} that you determined in Experiment 20, determine the quiescent drain current (either graphically or Equation 2a), and compare this value with that measured in Step 2. Your results should agree within 10 percent.
4. With your VOM or DMM, individually measure the following quiescent dc voltages: gate (V_G), source (V_S), drain (V_D), and drain-to-source (V_{DS}). Record your values in Table 21–2, and compare these values with the expected values calculated using Equations 3, 4, and 5 given in the "Useful Formulas' section of this experiment. Your results should agree within 10 percent. You should have measured essentially no gate voltage, since the gate current is very small due to the very high input impedance of the JFET.
5. Now measure the quiescent dc gate-to-source voltage (V_{GS}). How does this value compare with the source voltage measured in Step 4?

    These two voltages should be essentially the same, except that V_{GS} is a *negative* voltage while the source voltage is positive with respect to ground. In a self-biased circuit, the gate receives its bias voltage solely from the voltage developed across the source resistor. In addition, the JFET's quiescent gate-to-source voltage will always be negative unless the gate and source leads are shorted together.
6. With Equation 2b and the values of I_{DSS} and $V_{GS(off)}$ found in Experiment 20, plot on the blank graph provided for the transfer characteristic (transconductance) curve for the JFET you are using.
7. For the quiescent values of I_D and V_{GS} measured in Steps 2 and 5, plot the Q point on the graph. You should find that the Q point lies essentially on the characteristic curve.
8. Replace the 1-kΩ source resistors (R_S) with a 5-kΩ potentiometer. Connect your VOM and DMM as a voltmeter across the JFET's gate and source leads. Now vary the potentiometer slowly so that you can obtain approximately five simultaneous values for I_D and V_{GS}, recording them in Table 21–3.
9. Plot the data obtained in Step 8 on the graph. You should find that all the points lie essentially on the nonlinear, square-law transfer characteristic curve. This happens because the curve describes all the possible combinations of I_D and V_{GS} as I_D varies from zero to I_{DSS} while V_{GS} varies from zero to $V_{GS(off)}$. Furthermore, you should notice that changing R_S changes the

location of the Q point on the transconductance curve even though the supply voltage, I_{DSS}, and $V_{GS(off)}$ remain the same throughout.

WHAT YOU HAVE DONE

This experiment verified the voltages and currents in a JFET self-bias circuit. The gate-source voltage must be negative, but more positive than the JFET's gate-to-source cutoff voltage. The gate-to-source voltage is set up solely by the voltage drop of the circuit's source resistor. By varying the source resistor, a nonlinear square-law characteristic curve was obtained.

Name ______________________ Date __________

JFET SELF-BIAS

OBJECTIVES/PURPOSE:

SCHEMATIC DIAGRAM:

DATA FOR EXPERIMENT 21

TABLE 21–1

From Experiment 20	
I_{DSS}	mA
$V_{GS(off)}$	V
g_{m0}	μS

TABLE 21–2

Parameter	Measured Value	Expected Value	% Error
I_D			
V_G			
V_S			
V_D			
V_{DS}			
V_{GS}			

Name _____   Date _____

TABLE 21–3

V_{GS}	I_D

NOTES

Name _____   Date _____

DATA FOR EXPERIMENT 21

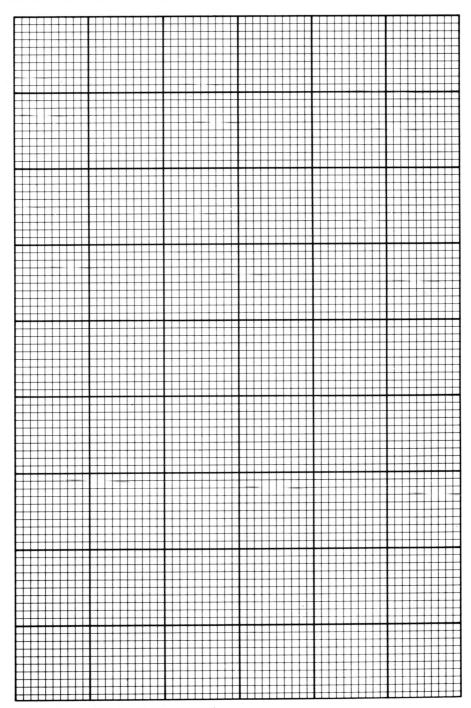

Name _____ Date _____

RESULTS AND CONCLUSIONS:

REVIEW QUESTIONS FOR EXPERIMENT 21

1. For an n-channel JFET used in a self-biased circuit,
   (a) the drain is more positive than the source
   (b) the source is more positive than the gate
   (c) the gate-to-source voltage is less negative than the $V_{GS(off)}$ of the JFET
   (d) all of the above   ( )

2. For the self-biased circuit of Figure 21–1 with $I_{DSS} = 10$ mA and $g_{m0} = 5000$ μS, the source current is approximately
   (a) 1 mA   (b) 2 mA   (c) 5 mA   (d) 10 mA   ( )

3. For the self-biased circuit of Figure 21–1 with $I_{DSS} = 10$ mA and $g_{m0} = 5000$ μS, the gate-to-source voltage is approximately
   (a) 1 V   (b) 2 V   (c) –1 V   (d) –2 V   ( )

4. For the self-biased circuit of Figure 21–1 with $I_{DSS} = 10$ mA and $g_{m0} = 5000$ μS, the JFET forward transconductance g_m at the Q point is approximately
   (a) 1000 μS   (b) 2500 μS   (c) 4000 μS   (d) 5000 μS   ( )

5. For the self-biased circuit of Figure 21–1 with $I_{DSS} = 10$ mA and $g_{m0} = 5000$ μS, the drain-to-source voltage is approximately
   (a) 2 V   (b) 3 V   (c) 5 V   (d) 10 V   ( )

198

22

THE DEPLETION-MODE MOSFET

PURPOSE AND BACKGROUND

The purpose of this experiment is to plot the transfer characteristics of a depletion-mode metal-oxide semiconductor field-effect transistor (MOSFET). The D-type MOSFET is sometimes called a *normally-on* or *depletion-enhancement* MOSFET as it can be operated with negative as well as positive gate-to-source voltages. Like the JFET, the D-mode MOSFET conducts drain current for V_{GS} between $V_{GS(off)}$ and 0 and can be biased in the same manner as a JFET. Because of the insulating layer between the gate and the channel, the MOSFET has a much higher input impedance than a JFET. Also because of this extremely high input impedance ($> 10^{12}$ Ω), MOSFETs are subject to destruction by static discharge and care must be taken when handling them.

Text References: 8–4, The Metal Oxide Semiconductor FET (MOSFET); 8–5, MOSFET Characteristics and Parameters; 8–6 MOSFET Biasing.

REQUIRED PARTS AND EQUIPMENT

☐ Two 10-kΩ resistors, 1/4 W
☐ 5-kΩ potentiometer
☐ 50-kΩ potentiometer
☐ 40673 n-channel MOSFET, or equivalent

☐ Two 0–15 V dc power supplies
☐ Two DMMs (preferred) or VOMs
☐ Breadboarding socket

USEFUL FORMULAS

Forward transconductance (at $V_{GS} = 0$)

$$(1) \quad g_{m0} = -\frac{2I_{DSS}}{V_{GS(off)}}$$

Quiescent dc drain current

$$(2) \quad I_D = I_{DSS}\left[1 - \frac{V_{GS}}{V_{GS(off)}}\right]^2$$

PROCEDURE

1. Wire the circuit shown in the schematic diagram of Figure 22–1. *Be very careful in handling the MOSFET.* Manufacturers normally package MOSFETs with all leads shorted together with a piece of wire or metal ring and the leads pressed into black conducting foam material. First insert the MOSFET on the breadboard and wire the remaining components. Then unwrap the shorting wire from all the leads. *Never insert a MOSFET to or remove a MOSFET from a circuit when the power is still on.*

2. Once you have removed the shorting wire or ring and checked the circuit for accuracy, apply power to the breadboard. Adjust the 5-kΩ potentiometer so that $V_{DS} = 10$ V.

3. Now adjust the 10-kΩ potentiometer so that $V_{GS} = -10$ V. Measure the corresponding drain current (I_D) and record this value in Table 22–1.

   The drain current may be zero as the gate-to-source voltage is more negative than $V_{GS(off)}$ for your particular MOSFET.

4. Adjust the 10-kΩ potentiometer for the remaining values of V_{GS} listed in Table 22–1. For each value of V_{GS}, measure the drain current and record its value.

5. From the data of Table 22–1, plot the characteristic, or transconductance curve (I_D vs. V_{GS}) for the MOSFET you are using on the blank graph provided for this purpose.

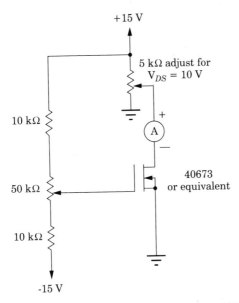

+15 V

5 kΩ adjust for
$V_{DS} = 10$ V

10 kΩ

+

A

—

40673
or equivalent

50 kΩ

10 kΩ

-15 V

FIGURE 22–1   *Schematic diagram of circuit.*

6. From the graph, determine I_{DSS} for your MOSFET at the point where $V_{GS} = 0$. Record its value in Table 22–2.
7. From the graph, determine $V_{GS(off)}$ for your MOSFET at the point where $I_D = 0$. Record its value in Table 22–2.
8. Using Equation 1 given in the "Useful Formulas" section of this experiment, determine the forward transconductance of your MOSFET, recording its value in Table 22–2.
9. When you are finished, first turn off the power to the breadboard. Then short all three leads of the MOSFET by wrapping them with a piece of wire before removing the MOSFET from the circuit.

WHAT YOU HAVE DONE

In this experiment you determined the transfer characteristic curve for a depletion-mode MOSFET. You were able to have gate-to-source voltages that were negative as well as positive. Once the curve was drawn, you determined the following parameters for your particular MOSFET: I_{DSS}, $V_{GS(off)}$, and g_{m0}. In addition, you learned how to properly connect a MOSFET to and remove a MOSFET from a circuit without damaging it.

NOTES

Name ______________________________ Date ____________

THE DEPLETION-MODE MOSFET

OBJECTIVES/PURPOSE:

SCHEMATIC DIAGRAM:

DATA FOR EXPERIMENT 22

TABLE 22–1

$V_{DS} = 10$ V	
V_{GS}	I_D (mA)
-10 V	
-8 V	
-6 V	
-4 V	
-2 V	
-1 V	
0 V	
1 V	
2 V	
4 V	
6 V	

TABLE 22–2

Parameter	Measured Value
I_{DSS}	mA
$V_{GS(\text{off})}$	V
g_{m0}	mS

Name _____ Date _____

DATA FOR EXPERIMENT 22

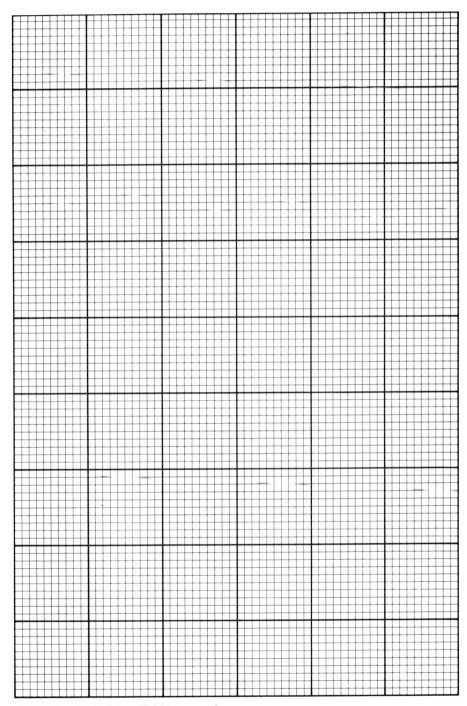

Name _____ Date _____

RESULTS AND CONCLUSIONS:

REVIEW QUESTIONS FOR EXPERIMENT 22

1. For an n-channel depletion-mode MOSFET to be properly biased, the gate-to-source voltage must be
   (a) negative
   (b) positive
   (c) zero
   (d) all of the above   (   )
2. Handling a MOSFET requires that
   (a) the shorting wire around all the leads must be removed before the device is connected to the circuit.
   (b) the MOSFET is never inserted or removed from a circuit when the power is on.   (   )
3. If $V_{GS} = -5$ V when $I_D = 0$, and $I_D = 10$ mA when $V_{GS} = 0$, the forward transconductance is
   (a) 2 mS   (b) 0.5 mS   (c) 4 mS   (d) 1 mS   (   )
4. For the circuit of Figure 22–1 if $V_{GS} = 1$ V, $V_{GS(off)} = -4$ V, and $I_{DSS} = 8$ mA, the drain current is
   (a) 0.5 mA   (b) 8 mA   (c) 10 mA   (d) 12.5 mA   (   )

206

23

THE ENHANCEMENT-MODE MOSFET

PURPOSE AND BACKGROUND

The purpose of this experiment is to plot the transfer characteristics of an enhancement-mode metal-oxide semiconductor field-effect transistor (MOSFET), which is sometimes called a *normally-off MOSFET.* When the gate-to-source voltage exceeds its *threshold voltage,* the MOSFET conducts and drain current flows. Because of the insulating layer between the gate and the channel, the enhancement-mode MOSFET has a much higher input impedance then a JFET. Also because of this extremely high input impedance ($> 10^{12}$ Ω), MOSFET are subject to destruction by static discharge and care must be taken when handling them.

Text References: 8–4, The Metal Oxide Semiconductor FET (MOSFET); 8–5, MOSFET Characteristics and Parameters; 8–6 MOSFET Biasing.

REQUIRED PARTS AND EQUIPMENT

Resistors (1/4 W):
- ☐ 1 kΩ
- ☐ 100 kΩ
- ☐ 1-MΩ potentiometer
- ☐ NTE 465 n-channel
  MOSFET, or equivalent

- ☐ 0–15 V dc power supply
- ☐ Two DMMs (preferred) or
  VOMs
- ☐ Breadboarding socket

USEFUL FORMULAS

Quiescent dc drain current

(1) $I_D = K \left[V_{GS} - V_{GS(\text{th})} \right]^2$

where K depends on the particular MOSFET and is determined
from the data sheet.

PROCEDURE

1. Wire the circuit shown in the schematic diagram of Figure 23–1.
   Be very careful in handling the MOSFET. Manufacturers normally package MOSFETs with all leads shorted together with
   a piece of wire or metal ring and the leads pressed into black

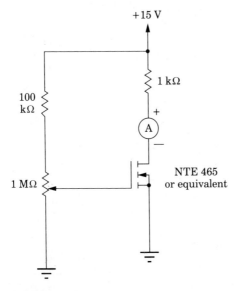

FIGURE 23–1 *Schematic diagram of circuit.*

conducting foam material. First insert the MOSFET on the breadboard and wire the remaining components. Then unwrap the shorting wire from all the leads. *Never insert a MOSFET to or remove a MOSFET from a circuit when the power is still on.*

2. Once you have removed the shorting wire or ring and checked the circuit for accuracy, apply power to the breadboard.

3. Now adjust the 1-MΩ potentiometer so that the gate-to-source voltage (V_{GS}) is zero. Slowly adjust this potentiometer and stop at the point where drain current (I_D) just starts to flow. Measure the gate-to-source voltage and record its value in Table 23–1. The gate-to-source voltage is the threshold voltage $V_{GS(th)}$.

4. Adjust the 1-mΩ potentiometer so that the gate-to-source voltage increases in 1-V steps up to 10 V. For each value of V_{GS}, measure the drain current and record both values in Table 23–1.

5. From the data of Table 23–1, plot the characteristic, or transconductance, curve (I_D vs. V_{GS}) for the enhancement-mode MOSFET you are using on the blank graph provided for this purpose.

6. When you are finished, first turn off the power to the breadboard. Then short all three leads of the MOSFET by wrapping them with a piece of wire before removing the MOSFET from the circuit.

WHAT YOU HAVE DONE

In this experiment you determined the transfer characteristic curve for an enhancement-mode MOSFET. Drain current conducts when the gate-to-source voltage exceeds its threshold voltage and increases in a parabolic (square law) fashion. In addition, you learned how to properly connect a MOSFET to and remove a MOSFET from a circuit without damaging it.

NOTES

Name _____   Date _____

THE ENHANCEMENT-MODE MOSFET

OBJECTIVES/PURPOSE:

SCHEMATIC DIAGRAM:

211

DATA FOR EXPERIMENT 23

TABLE 23–1

V_{GS}	I_D (mA)

DATA FOR EXPERIMENT 23

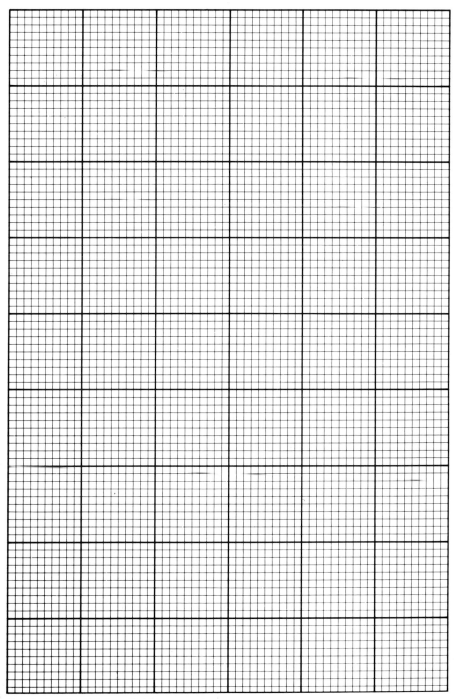

RESULTS AND CONCLUSIONS:

REVIEW QUESTIONS FOR EXPERIMENT 23

1. For an n-channel depletion-mode MOSFET to be properly biased, the gate-to-source voltage must be
   (a) negative
   (b) positive
   (c) zero
   (d) more positive than its threshold voltage   ( )
2. Handling a MOSFET requires that
   (a) the shorting wire around all the leads must be removed
       before the device is connected to the circuit.
   (b) the MOSFET is never inserted or removed from a circuit
       when the power is on.   ( )
3. If I_D = 3 mA when V_{GS} = 5 V and $V_{GS(th)}$ = 2 V, the drain
   current when V_{GS} = 7 V is
   (a) 5 mA   (b) 8.3 mA   (c) 7 mA   (d) 3 mA   ( )
4. If I_D = 4 mA when V_{GS} = 4 V and $V_{GS(th)}$ = 3 V, the drain
   current when V_{GS} = 2 V is
   (a) 2 mA   (b) 3 mA   (c) 4 mA   (d) 0 mA   ( )

214

24

VMOSFET RELAY DRIVER

PURPOSE AND BACKGROUND

The purpose of this experiment is to demonstrate the use of a VMOSFET as a driver interface for a 12-V relay, providing isolation between one circuit and another. Because of its construction, the VMOSFET is sometimes called a *power MOSFET* and is capable of handling higher currents than regular enhancement-mode MOSFETs. Like other MOSFETs, the VMOSFET has a much higher input impedance than a JFET. Also because of this extremely high input impedance ($> 10^{12}$ Ω), VMOSFETs are subject to destruction by static discharge, so care must be taken when handling them.

Text References: 8–4, The Metal Oxide Semiconductor FET (MOSFET); 8–5, MOSFET Characteristics and Parameters; 8–6 MOSFET Biasing.

REQUIRED PARTS AND EQUIPMENT

Resistors (1/2 W):
- ☐ 100 kΩ
- ☐ 1 MΩ
- ☐ VK67AK n-channel MOSFET, or equivalent

- ☐ 0–15 V dc power supply
- ☐ 12-V relay
- ☐ SPDT switch
- ☐ DMM (preferred) or VOM
- ☐ Breadboarding socket

215

PROCEDURE

1. Wire the circuit shown in the schematic diagram of Figure 24–1. *Be very careful in handling the VMOSFET.* Manufacturers normally package VMOSFETs with all leads shorted together with a piece of wire or metal ring and the leads pressed into black conducting foam material. First insert the VMOSFET on the breadboard and wire the remaining components. Then unwrap the shorting wire from all the leads. *Never insert a VMOSFET to or remove a VMOSFET from a circuit when the power is still on.*

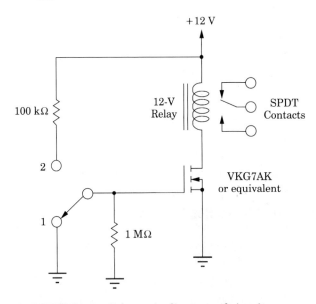

FIGURE 24–1 *Schematic diagram of circuit.*

2. Once you have removed the shorting wire or ring and checked the circuit for accuracy, apply power to the breadboard.
3. First place the switch in position 1. Measure V_{GS} and V_{DS}, recording both values in Table 24–1.

    Since $V_{GS} = 0$, the VMOSFET is cutoff, such that $V_{GS} < V_{GS(th)}$. Since $I_D = 0$, there is no voltage drop across the relay coil and the relay armature is not energized.
4. Then place the switch in position 2. Measure V_{GS} and V_{DS}, recording both values in Table 24–1.

    Since $V_{GS} > V_{GS(th)}$, the VMOSFET conducts. The relay armature pulls in. The relay contacts are then able to control the switching of another circuit and is electrically isolated.
5. When you are finished, first turn off the power to the breadboard. Then short all three leads of the VMOSFET by wrapping them with a piece of wire before removing the VMOSFET from the circuit.

WHAT YOU HAVE DONE

In this experiment you controlled a relay using a VMOSFET interface driver. When the VMOSFET is cutoff, the relay is not energized. When a positive voltage is applied to the gate lead, the relay armature pulls in.

NOTES

Name ______________________ Date ____________

VMOSFET RELAY DRIVER

OBJECTIVES/PURPOSE:

SCHEMATIC DIAGRAM:

Name _____   Date _____

DATA FOR EXPERIMENT 24

TABLE 24–1

Switch Position	V_{GS}	V_{DS}	Relay Energized?
1			
2			

Name _____   Date _____

RESULTS AND CONCLUSIONS:

REVIEW QUESTIONS FOR EXPERIMENT 24

1. Handling a VMOSFET requires that
   (a) the shorting wire around all the leads must be removed before the device is connected to the circuit.
   (b) the VMOSFET is never inserted or removed from a circuit when the power is on.   (   )
2. For the relay driver circuit of Figure 24–1, when the switch is in position 1, the VMOSFET is at
   (a) saturation   (b) cutoff   (   )
3. For the relay driver circuit of Figure 24–1, when the switch is in position 1, the relay armature is
   (a) energized   (b) not energized   (   )
4. When the relay is energized in the circuit of Figure 24–1,
   (a) $V_{GS} > V_{GS(\text{th})}$ (b) $V_{GS} < V_{GS(\text{th})}$ (   )

NOTES

25

THE COMMON-SOURCE
AMPLIFIER

PURPOSE AND BACKGROUND

The purposes of this experiment are (1) to demonstrate the operation and characteristics of a self-biased common-source amplifier and (2) to investigate what influences its voltage gain by using the JFET parameters measured in Experiment 20. The common-source amplifier is characterized by application of the amplifier input signal to the gate lead while its output is taken from the drain, a condition that always gives a 180° phase shift.

Text References: 9–3, Common-Source Amplifiers.

REQUIRED PARTS AND EQUIPMENT

Resistors (1/4 W):
- ☐ Two 1 kΩ
- ☐ 4.7 kΩ
- ☐ 100 kΩ

Capacitors (25 V):
- ☐ Two 2.2 μF
- ☐ 100 μF

- ☐ MPF102 n-channel JFET
- ☐ 0–15 V dc power supply
- ☐ Signal generator
- ☐ VOM or DMM (preferred)
- ☐ Dual trace oscilloscope
- ☐ Breadboarding socket

USEFUL FORMULAS

Voltage gain

(1) $A_v = \dfrac{v_{\text{out}}}{v_{\text{in}}}$

(2) $A_v = g_m R_d$ (source resistor bypassed)

(3) $A_v = \dfrac{g_m R_d}{1 + g_m R_S}$ (source resistor not bypassed)

where $R_d = R_D \| R_L$

JFET dc gate-to-source cutoff voltage

(4) $V_{GS(\text{off})} = -\dfrac{2I_{DSS}}{g_{m0}}$

Quiescent dc drain (source) current

(5) $I_D = 2I_{DSS}\left[\dfrac{(R_S g_{m0} + 1) - (2R_S g_{m0} + 1)^{1/2}}{(R_S g_{m0})^2} \right]$

Quiescent dc gate-to-source voltage

(6) $V_S = I_D R_S$

(7) $= -V_{GS}$ so that $V_G \simeq 0$

Quiescent dc drain-to-source voltage

(8) $V_{DS} = V_{DD} - I_D(R_D + R_S)$

JFET forward transconductance at Q point

(9) $g_m = g_{m0}\left[1 - \dfrac{V_{GS}}{V_{GS(\text{off})}} \right]$

PROCEDURE

1. Wire the circuit shown in Figure 25–1, omitting the signal generator and the power supply. Enter the values of I_{DSS}, $V_{GS(\text{off})}$, and g_{m0} obtained in Experiment 14 in Table 25–1.
2. After you have checked all connections, apply only the 15-V supply voltage to the breadboard. With your VOM or DMM, measure the JFET's quiescent drain current I_D and gate-to-source voltage V_{GS}, recording your values in Table 25–2. Based on the values for g_{m0} and I_{DSS} that you measured for the JFET in Experiment 20, calculate the JFET's transconductance g_m at this quiescent point, and record your result in Table 25–2. In all cases, compare your measured values with what you would expect to measure.

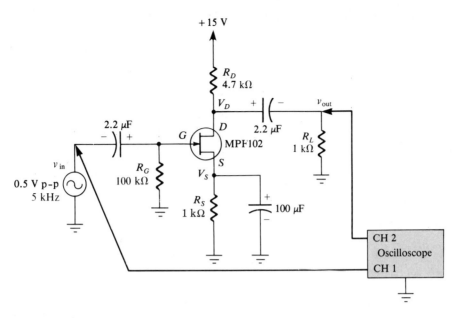

FIGURE 25–1 *Schematic diagram of circuit.*

3. Connect Channel 1 of your oscilloscope at the input to the amplifier (v_{in}) and Channel 2 to the 1-kΩ load resistor (v_{out}). Then
connect the signal generator to the circuit as shown in Figure
25–1, and adjust the sine wave output level of the generator at
0.5 V *peak-to-peak* at a frequency level of 5 kHz.

   Note that the output signal level (v_{out}) is *greater* than the
input level (v_{in}). In addition, v_{out} is inverted, or 180° out-of-
phase, with respect to the input. These points are two major
characteristics of a common-source amplifier. In order to observe
the phase shift, you must display both signals *simultaneously*
on the oscilloscope; otherwise, you will not see any phase shift.

4. With an oscilloscope, measure the ac peak-to-peak voltage at
the JFET's source lead. Even at the oscilloscope's highest input
sensitivity, you should measure virtually no ac voltage at this
point. The 100-μF bypass capacitor, in parallel with R_S, serves
essentially as a short circuit path to ground, since its reactance
at 5 kHz is very small compared with the 1-kΩ resistance. Consequently, the source lead is effectively at *ac ground*.

5. Calculate the expected voltage gain from Equation 2 using the
transconductance value determined in Step 2, and record the
value in Table 25–2. Now measure the actual circuit voltage
gain by dividing the peak-to-peak output voltage v_{out} by the
peak-to-peak input voltage v_{in}, recording your result in Table
25–2.

6. Now remove R_L. You should observe that the output voltage level increases. It does so because the load resistance affects the voltage gain of the amplifier stage. As in Step 5, experimentally determine the voltage gain by measuring v_{out} and v_{in}, comparing your measured result with the expected value (Equation 2, with $R_d = R_D$). Record your results in Table 25–2.

7. In this final step, also remove the 100-μF source bypass capacitor from the circuit. Note that the output voltage decreases tremendously from that of Step 6. As in the previous two steps, experimentally determine the voltage gain by measuring v_{out} and v_{in}, comparing your results with the expected value (Equation 3, with $R_d = R_D$). Record your results in Table 25–2.

From the results in Table 25–2, you should now understand how both the source and load resistances affect the voltage gain of a common-source amplifier.

WHAT YOU HAVE DONE

This experiment demonstrated the operation and characteristics of a self-biased common-source amplifier, which has 180° phase shift. Here, the input signal is applied to the JFET's gate lead, while the output signal is taken from the drain lead. The experiment also showed how both the load resistance and emitter bypass capacitor influenced the circuit's voltage gain.

Name ______________________ Date ____________

THE COMMON-SOURCE AMPLIFIER

OBJECTIVES/PURPOSE:

SCHEMATIC DIAGRAM:

DATA FOR EXPERIMENT 25

TABLE 25–1

From Experiment 20:			
g_{m0} = $\qquad\qquad$ I_{DSS} =			
Parameter	Measured Value	Expected Value	% Error
I_D			
V_{GS}			
g_m	■		■

TABLE 25–2

Condition	v_{in}	v_{out}	Measured Gain	Expected Gain	% Error
Normal circuit (Step 5)					
No load (Step 6)					
No bypass capacitor and no load (Step 7)					

Name _____ Date _____

RESULTS AND CONCLUSIONS:

REVIEW QUESTIONS FOR EXPERIMENT 25

1. For the circuit of Figure 25–1 with I_{DSS} = 10 mA and g_{m0} = 5000 μS, the voltage gain from gate to drain is approximately
   (a) 0.6   (b) 0.7   (c) 2   (d) 4   (   )
2. The signal at the drain is out-of-phase with the gate by
   (a) 0°   (b) 45°   (c) 90°   (d) 180°   (   )
3. If the source bypass capacitor in Figure 25–1 is removed, the voltage gain will
   (a) increase   (b) decrease
   (c) remain essentially the same   (   )
4. If the load resistor R_L in the circuit of Figure 25–1 is increased, the voltage gain will
   (a) increase   (b) decrease
   (c) remain essentially the same   (   )
5. The common-source amplifier of Figure 25–1 is similar in operation to a bipolar transistor
   (a) common-base amplifier   (b) common-collector amplifier
   (c) common-emitter amplifier   (d) emitter-follower   (   )

NOTES

26

THE COMMON-DRAIN
AMPLIFIER
(SOURCE-FOLLOWER)

PURPOSE AND BACKGROUND

The purposes of this experiment are (1) to demonstrate the operation and characteristics of a self-biased common-drain amplifier, and (2) using the JFET parameters measured in Experiment 20, to investigate what influences its voltage gain. The common-drain amplifier, often referred to as a *source-follower,* is characterized by application of the input signal to the gate lead while the output is taken from the source. The output signal is never larger than the input but is always in-phase with the input. Consequently, the output *follows* the input.

Text Reference: 9–4, Common-Drain Amplifiers.

REQUIRED PARTS AND EQUIPMENT

Resistors (1/4 W):
- ☐ Two 1 kΩ
- ☐ 100 kΩ
- ☐ Two 2.2-μF capacitors, 25 V
- ☐ MPF102 n-channel JFET

- ☐ 0–15 V dc power supply
- ☐ Signal generator
- ☐ VOM or DMM (preferred)
- ☐ Dual trace oscilloscope
- ☐ Breadboarding socket

USEFUL FORMULAS

Voltage gain

$$(1) \ A_v = \frac{v_{\text{out}}}{v_{\text{in}}}$$

$$(2) \ A_v = \frac{g_m R_s}{g_m R_s + 1}$$

where $R_s = R_S \| R_L$

JFET dc gate-to-source cutoff voltage

$$(3) \ V_{GS(\text{off})} = \frac{2I_{DSS}}{g_{m0}}$$

Quiescent dc drain (source) current

$$(4) \ I_D = 2I_{DSS} \left[\frac{(R_S g_{m0} + 1) - (2R_S g_{m0} + 1)^{1/2}}{(R_S g_{m0})^2} \right]$$

Quiescent dc gate-to source voltage

$$(5) \ V_S = I_D R_S$$

$$(6) \quad = -V_{GS} \qquad \text{so that } V_G \simeq 0$$

Quiescent dc drain-to-source voltage

$$(7) \ V_{DS} = V_{DD} - I_D R_S$$

JFET forward transconductance at Q point

$$(8) \ g_m = g_{m0} \left[1 - \frac{V_{GS}}{V_{GS(\text{off})}} \right]$$

PROCEDURE

1. Wire the circuit shown in Figure 26–1, omitting the signal generator and the power supply.
2. After you have checked all connections, apply only the 15-V supply voltage to the breadboard. With your VOM and DMM, measure the JFET's quiescent drain current I_D and gate-to-source voltage V_{GS}, recording your values in Table 26–1. Based on the values for g_{m0} and I_{DSS} that you measured for the same JFET in Experiment 20, calculate the JFET's transconductance g_m at this quiescent point, and record your result in Table 26–1. In all cases, compare your measured values with what you would expect to measure.

3. Connect Channel 1 of your oscilloscope to the amplifier's input (v_{in}) and Channel 2 to the 1-kΩ load resistor (v_{out}). Then connect the signal generator to the circuit as shown in Figure 26–1, and adjust the sine wave output level of the generator at 1 V *peak-to-peak* at a frequency of 5 kHz.

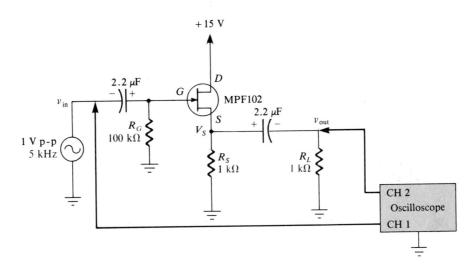

FIGURE 26–1 *Schematic diagram of circuit.*

Note that the output signal level (v_{out}) is less than the input level (v_{in}). In addition, v_{out} is in-phase with the input. These points are two major characteristics of a common-drain amplifier.

4. Calculate the voltage gain (Equation 2) using the transconductance value determined in Step 2, and record the value in Table 26–2. Now measure the actual circuit voltage gain by dividing the peak-to-peak output voltage v_{out} by the peak-to-peak input voltage v_{in} (Equation 1), recording your result in Table 26–2.

5. Now remove R_L. Observe that the output voltage level increases somewhat. It does so because the load resistance affects the voltage gain of the amplifier stage. As in Step 4, experimentally determine the voltage gain by measuring v_{out} and v_{in}, comparing your measured result with the expected value. In this case, $R_s = R_S$. Record your results in Table 26–2.

From the results in Table 26–2, you should now understand how the load resistance affects the voltage gain of a common-drain amplifier.

WHAT YOU HAVE DONE

This experiment demonstrated the operation and characteristics of a self-biased common-drain amplifier, or *source follower*, which has no phase shift. Here, the input signal is applied to the JFET's gate lead, while the output signal is taken from the source lead. The experiment also showed how the load resistance influenced the circuit's voltage gain.

Name ______________________________ Date ____________

THE COMMON-DRAIN AMPLIFIER (SOURCE-FOLLOWER)

OBJECTIVES/PURPOSE:

SCHEMATIC DIAGRAM:

DATA FOR EXPERIMENT 26

TABLE 26–1

From Experiment 20:			
g_{m0} = I_{DSS} =			
Parameter	Measured Value	Expected Value	% Error
I_D			
V_{GS}			
g_m	▓▓▓▓▓▓▓		▓▓▓▓▓▓▓

TABLE 26–2

Condition	v_{in}	v_{out}	Measured Gain	Expected Gain	% Error
Normal circuit (Step 4)					
No load (Step 5)					

Name _____   Date _____

RESULTS AND CONCLUSIONS:

REVIEW QUESTIONS FOR EXPERIMENT 26

1. For the circuit of Figure 26–1 with I_{DSS} = 10 mA and g_{m0} = 5000 μS, the voltage gain from gate to source is approximately
   (a) 0.56   **(b)** 0.71   **(c)** 0.83   **(d)** 1   (   )
2. The signal at the source is out-of-phase with the gate by
   (a) 0°   **(b)** 45°   **(c)** 90°   **(d)** 180°   (   )
3. If the JFET forward transconductance in the circuit of Figure 26–1 is increased, the voltage gain will
   (a) increase   **(b)** decrease
   (c) remain essentially the same   (   )
4. If the load resistor R_L in the circuit of Figure 26–1 is decreased, the voltage gain will
   (a) increase   **(b)** decrease
   (c) remain essentially the same   (   )
5. The common-drain amplifier is similar in operation to a bipolar transistor
   (a) common-base amplifier   **(b)** common-collector amplifier
   (c) common-emitter amplifier   (   )

237

NOTES

27

AMPLIFIER LOW-FREQUENCY RESPONSE

PURPOSE AND BACKGROUND

The purpose of this experiment is to demonstrate the factors that contribute to the low-frequency response of a common-emitter transistor amplifier. The low-frequency response of a typical common-emitter amplifier is determined in part by the input and output coupling capacitors and the emitter bypass capacitor. The result is essentially a combination of three high-pass filter networks that allow signals having frequencies greater than the cutoff frequency of the dominant network to pass through while attenuating all others. Although the cutoff frequencies associated with these three paths can be made equal, such is rarely the case.

This experiment examines individually the effect of each capacitor on the common-emitter amplifier's low-frequency response. In all cases, the values of the capacitors are intentionally made abnormally small in order to allow the frequency response to be easily measured.

Text References: 10–1, General Concepts; 10–2, Decibels; 10–3, Low-Frequency Amplifier Response; 10–6 Total Amplifier Response.

REQUIRED PARTS AND EQUIPMENT

Resistors (1/4 W):
- ☐ 150 Ω
- ☐ Two 2.7 kΩ
- ☐ 3.9 kΩ
- ☐ 47 kΩ
- ☐ 100 kΩ

Capacitors:
- ☐ Two 0.033 μF
- ☐ 1 μF
- ☐ Two 2.2 μF, 25 V
- ☐ 10 μF, 25 V

- ☐ 2N3904 npn silicon transistor
- ☐ 0–15 V dc power supply
- ☐ VOM or DMM (preferred)
- ☐ Signal generator
- ☐ Dual trace oscilloscope
- ☐ Breadboarding socket

USEFUL FORMULAS

Quiescent dc base voltage

$$(1)\ V_B = \left(\frac{R_2}{R_1 + R_2}\right)V_{CC}$$

Quiescent dc emitter voltage

$$(2)\ V_E = V_B - V_{BE}$$

Quiescent dc emitter current

$$(3)\ I_E = \frac{V_E}{R_{E1} + R_{E2}}$$

Transistor ac emitter resistance (at normal room temperature)

$$(4)\ r_e \cong \frac{25\ \text{mV}}{I_E}$$

Amplifier ac input impedance

$$(5)\ R_{\text{in}} = R_1 \parallel R_2 \parallel \beta(r_e + R_{E1})$$

$$(6)\ A_v = \frac{v_{\text{out}}}{v_{\text{in}}}$$

$$(7)\ A_v = \frac{R_C \parallel R_L}{R_{E1} + r_e}$$

Decibel voltage gain

$$(8)\ \text{dB} = 20 \log(A_v)$$

Frequency response due to input coupling capacitor C_1

$$(9)\ f_1 = \frac{1}{2\pi C_1(R_{\text{in}} + R_G)}$$

Frequency response due to emitter bypass capacitor C_2

$$(10) \quad f_2 = \frac{1}{2\pi C_2 \left[\left(\dfrac{R_1 \parallel R_2 \parallel R_G}{\beta} + R_{E1} + r_e \right) \parallel R_{E2} \right]}$$

Frequency response due to output coupling capacitor C_3

$$(11) \quad f_3 = \frac{1}{2\pi C_3 (R_C + R_L)}$$

PROCEDURE

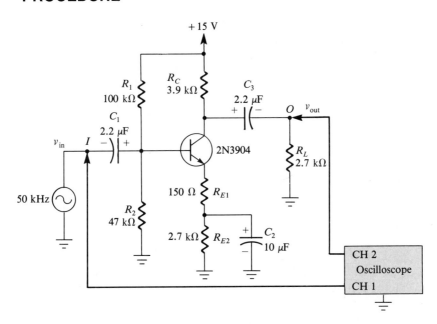

FIGURE 27–1 *Schematic diagram of circuit.*

1. Wire the circuit shown in Figure 27–1, omitting the signal generator and the power supply.
2. After you have checked all connections, apply the 15-V supply voltage to the breadboard. With a VOM or DMM, measure the transistor's quiescent dc emitter voltage with respect to ground. From this value, determine the transistor's ac internal emitter resistance r_e (Equation 4). Then determine the expected midband voltage gain of the amplifier in decibels (Equations 7 and 8). Record these values in Table 27–1.
3. Connect Channel 1 of your oscilloscope to point I (v_{in}) and Channel 2 to point O (v_{out}). Then connect the signal generator to the circuit as shown in Figure 27–1, and adjust the sine wave output level of the generator at a frequency of 50 kHz so that the

241

peak-to-peak output voltage of the amplifier spans 7.1 vertical divisions when Channel 2 is set at a sensitivity of 0.5 V/division. Measure the peak-to-peak input voltage level, and determine the amplifier's dB voltage gain (Equation 8). Record this value in Table 27–1.

4. In order to determine the amplifier's low-frequency 3-dB point due solely to the effects of the input coupling capacitor C_1, replace C_1 (2.2 μF) with a 0.033-μF capacitor. Adjust the sine wave output level of the generator at a frequency of 50 kHz so that the peak-to-peak output voltage of the amplifier spans 7.1 vertical divisions when Channel 2 is set at a sensitivity of 0.5 V/division. Then slowly reduce the input frequency until the peak-to-peak output voltage drops to 5 vertical divisions. This reduction in output voltage is 5/7.1, or 0.707, which is equivalent to -3 dB. Using your oscilloscope, measure the frequency at which this value occurs, and record this frequency (f_1) in Table 27–2 along with the expected value (Equation 9) for comparison, assuming a typical beta of 150 for the 2N3904 transistors. Then replace C_1 with a 2.2-μF capacitor. You should observe that the output signal becomes smaller as the input frequency is reduced.[*]

5. In order to determine the amplifier's low-frequency 3-dB point due solely to the effects of the emitter bypass capacitor C_2, replace C_2 (10 μF)with a 1-μF capacitor. Again adjust the sine wave output level of the generator at a frequency of 50 kHz so that the peak-to-peak output voltage of the amplifier spans 7.1 vertical divisions when Channel 2 is set at a sensitivity of 0.5 V/division. Then reduce the input frequency until the peak-to-peak output voltage drops to 5 vertical divisions. From your oscilloscope, measure the frequency at which this value occurs (f_2), and record this frequency in Table 27–2 along with the expected value for comparison (Equation 10). Again assume that beta is typically 150. Then replace C_2 with a 10-μF capacitor. You should again observe that the output signal becomes smaller as in the input frequency is reduced.

6. In order to determine the amplifier's low-frequency 3-dB point due solely to the effects of the output coupling capacitor C_3, replace C_3 (2.2 μF) with a 0.033-μF capacitor. Again adjust the sine wave output level of the generator at a frequency of 50 kHz so that the peak-to-peak output voltage of the amplifier spans 7.1 vertical divisions when Channel 2 is set at a sensitivity of 0.5 V/division. Then reduce the input frequency until the peak-to-peak output voltage drops to 5 vertical divisions. From your

---

[*]For most laboratory signal generators, R_G = 50 Ω. If you are not sure of yours, consult the generator's user manual or ask your instructor.

oscilloscope, measure the frequency at which this value occurs (f_3), and record this frequency in Table 27–2 along with the expected value for comparison (Equation 11). You should observe that the output signal becomes smaller as the input frequency is reduced.

WHAT YOU HAVE DONE

This experiment demonstrated the factors that control the low-frequency response of a small-signal, common-emitter amplifier. You determined the amplifier's midband voltage gain, the critical frequencies of the input R_C circuit, the emitter bypass circuit, and the output R_C circuit.

NOTES

Name ______________________________ Date ____________

AMPLIFIER LOW-FREQUENCY RESPONSE

OBJECTIVES/PURPOSE:

SCHEMATIC DIAGRAM:

Name _____   Date _____

DATA FOR EXPERIMENT 27

TABLE 27–1 *Amplifier midband response.*

Parameter	Value
V_E (measured)	
r_e (measured)	
v_{in}	
v_{out}	
A_V(dB) (measured)	
A_V(dB) (expected)	
% error	

TABLE 27–2 *Amplifier low-frequency response (critical frequencies).*

Frequency	Measured	Expected	% Error
f_1			
f_2			
f_3			

Name _____ Date _____

RESULTS AND CONCLUSIONS:

REVIEW QUESTIONS FOR EXPERIMENT 27

1. The midband decibel gain for the circuit of Figure 27–1 is approximately
   (a) 9.5 dB (b) 19.5 dB (c) 23.6 dB (d) 27.4 dB   (   )
2. The low-frequency response of the amplifier of Figure 27–1 is controlled by
   (a) capacitor C_1 (b) capacitor C_2
   (c) capacitor C_3 (d) all of the above   (   )
3. Assuming a signal source impedance of 50 Ω and a β of 100 for the circuit for Figure 27–1, the critical frequency due solely to C_1 is approximately
   (a) 1 Hz (b) 3 Hz (c) 6 Hz (d) 43 Hz   (   )
4. Assuming a signal source impedance of 50 Ω and a β of 100 for the circuit of Figure 27–1, the critical frequency due solely to C_2 is approximately
   (a) 2 Hz (b) 6 Hz (c) 50 Hz (d) 100 Hz   (   )
5. Assuming a signal source impedance of 50 Ω and a β of 100 for the circuit of Figure 27–1, the critical frequency due solely to C_3 is approximately
   (a) 11 Hz (b) 19 Hz (c) 27 Hz (d) 45 Hz   (   )

247

NOTES

28

THE SILICON-CONTROLLED RECTIFIER

PURPOSE AND BACKGROUND

The purposes of this experiment are to demonstrate (1) how to test a silicon-controlled rectifier (SCR) with an ohmmeter and (2) how to use an SCR to create a half-wave, variable-resistance, phase-control circuit. Essentially the equivalent of a pnp and an npn transistor connected together, the SCR is a special type of diode that will not conduct current from anode to cathode unless a sufficient amount of current is applied to its gate terminal. Once an SCR has been turned on by such a gate pulse, it continues to conduct even if the gate pulse is removed. The only way to turn off an SCR is to remove the voltage from the anode (that is, zero volts).

This experiment demonstrates how an SCR is used to vary the power dissipated by a resistive load by varying the SCR's conduction angle between 0° and 90°. Such a circuit can be used as a light dimmer or a variable-speed control for an electric motor, such as a hand drill.

Text References: 11–2, The Silicon-Controlled Rectifier (SCR); 11–3, SCR Applications.

REQUIRED PARTS AND EQUIPMENT

Resistors:
- [ ] 470 Ω, 1/4W
- [ ] 2.2 kΩ, 1/2 W
- [ ] 5-kΩ potentiometer, or 10-turn "trimpot"
- [ ] 1N4001 silicon rectifier diode

- [ ] 2N6397 SCR, 200 V, 5 A
- [ ] 12.6-V rms secondary transformer
- [ ] Dual trace oscilloscope
- [ ] VOM or DMM (preferred)
- [ ] Breadboarding socket

USEFUL FORMULA

rms power dissipated by a resistor

$$P = \frac{V_{rms}^2}{R}$$

PROCEDURE

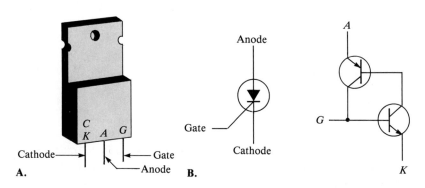

FIGURE 28–1 *Pin configuration and diode junction representation of SCR.*

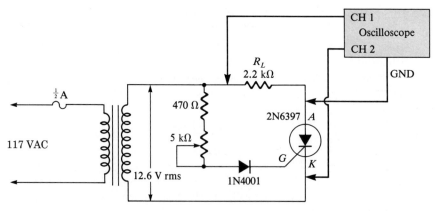

FIGURE 28–2 *Schematic diagram of circuit.*

1. As with diodes and transistors (see Experiments 1 and 8), a VOM can be used to check quickly whether an SCR is good or bad. Unless they have a specific function for this purpose, *DMMs are not able to perform this test properly.* Refer to Experiment 8 and reread how to determine the polarity of your ohmmeter's leads.

2. Using a 2N6397 or equivalent SCR, whose pin diagram and diode junction representation are shown in Figure 28–1, connect the ohmmeter's positive lead to the SCR's anode lead, while the ohmmeter's negative lead is connected to the cathode lead. Measure the resistance, recording this value in Table 28–1.

3. Reverse the meter's leads so that the positive lead is connected to the cathode while the ohmmeter's negative lead is connected to the anode lead. Measure the resistance and record this value in Table 28–1.

   The meter readings in both cases should be about the same.

4. Now connect the meter's positive lead to the gate lead, and the negative lead to the cathode lead. Measure the resistance and record this value in Table 28–1.

5. Reverse the meter leads. Measure the resistance and record this value in Table 28–1.

6. Again connect the meter's positive lead to the gate lead, but now connect the meter's negative lead to the anode. Measure the resistance and record this value in Table 28–1.

7. Reverse the meter leads. Measure the resistance and record this value in Table 28–1.

8. In Steps 2, 3, 6, and 7, the measured resistance is generally greater than 1 MΩ. In Step 4, you forward biased the gate-cathode junction, while this junction was reverse biased in Step 5. Consequently, the resistance reading obtained in Step 4 should be less than in Step 5, similar to that of an ordinary diode.

9. Wire the circuit shown in the schematic diagram of Figure 28–2. Initially set the 5-kΩ potentiometer to approximately 2 kΩ. In addition, set your oscillscope to the following approximate settings:

   Channel 1: 10 V/division, dc coupling
   Channel 2: 20 V/division, dc coupling
   Time base: 5 ms/division

   The ground lead of the oscilloscope probe should be connected to the *anode* terminal of the SCR.

10. Apply power to the breadboard. On Channels 1 and 2 you should see approximately three cycles of the voltage across the 2.2-kΩ load resistor and a cathode-to-anode voltage waveform like that shown in Figure 28–3. Because of the manner in

which the oscilloscope probes are connected to the circuit, the voltage across the SCR (Channel 2) is displayed *inverted* from what you would normally expect to see.

11. Change the oscilloscope's time base to 1 ms/division. Using the voltage waveforms shown in Figure 28–4 as a guide, measure the SCR's conduction angle for various settings of the 5-kΩ potentiometer. Based on the time divisions in Figure 28–4, the SCR's conduction angle (α) in degrees is computed as

$$\alpha = \frac{t_1}{2t_2} \times 360°$$

For each setting of the potentiometer, measure t_1 and t_2. Also, use an ac voltmeter to measure the rms voltage across the

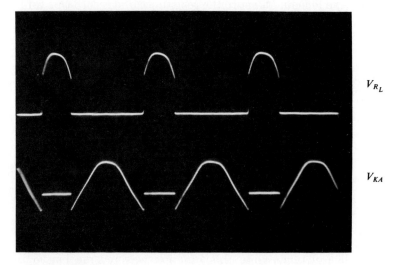

FIGURE 28–3

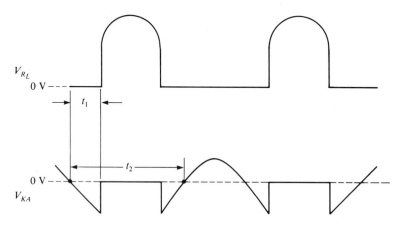

FIGURE 28–4

2.2-kΩ load resistor. Calculate the power dissipated by the load at each setting. You should attempt about 10 different settings. Record your values in Table 28–2.

12. On the blank graph provided for this purpose, plot the rms power dissipated by the load versus the SCR's conduction angle. What do you conclude from the graph?

From your graph, you should see that the load's dissipated power decreases with increasing conduction angle.

WHAT YOU HAVE DONE

This experiment demonstrated how to properly test a silicon-controlled rectifier (SCR) with an ohmmeter. In addition, this experiment demonstrated the operation of a half-wave phase-control circuit whose SCR's conduction angle is controlled from 0° to 90° by a variable resistor. Data was taken so that a graph of the power dissipated by the load vs. the SCR's conduction angle could be made.

NOTES

Name ______________________________ Date __________

THE SILICON-CONTROLLED RECTIFIER

OBJECTIVES/PURPOSE:

SCHEMATIC DIAGRAM:

DATA FOR EXPERIMENT 28

TABLE 28–1 *SCR resistance tests.*

Step Number	Ohmmeter Leads +	Ohmmeter Leads −	Resistance
2	anode	cathode	
3	cathode	anode	
4	gate	cathode	
5	cathode	gate	
6	gate	anode	
7	anode	gate	

Name _____   Date _____

TABLE 28–2 *SCR variable phase control*

t_1	t_2	α	V_L(rms)	Load Power

NOTES

Name _____ Date _____

DATA FOR EXPERIMENT 28

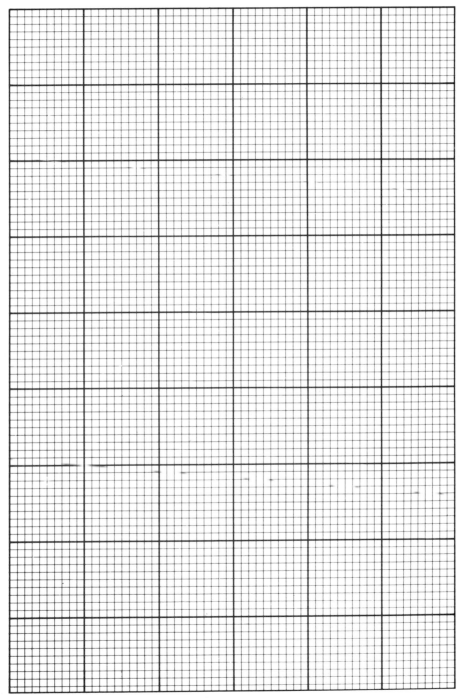

Name _____ Date _____

RESULTS AND CONCLUSIONS:

REVIEW QUESTIONS FOR EXPERIMENT 28

1. The equivalent circuit representation for an SCR uses
    (a) an npn transistor and a diode
    (b) a pnp transistor and a diode
    (c) two transistors of the same type
    (d) an npn and a pnp transistor   (   )
2. If the ohmmeter's positive lead is connected to the SCR's anode
    while the negative lead is connected to the cathode, for a "good"
    SCR the ohmmeter will read
    (a) 0
    (b) a very low resistance
    (c) typically several thousand ohms
    (d) a nearly infinite resistance   (   )
3. If the ohmmeter's positive lead is connected to the SCR's anode
    while the negative lead is connected to the gate, for a "good"
    SCR the ohmmeter will read
    (a) 0
    (b) a very low resistance
    (c) typically several thousand ohms
    (d) a nearly infinite resistance   (   )
4. For the circuit of Figure 28–1, once the SCR conducts, the way
    to turn off the SCR is to
    (a) reverse bias the gate
    (b) forward bias the gate
    (c) remove, or ground, the anode voltage
    (d) ground the gate lead   (   )
5. As the SCR's conduction angle is increased from 0° to 90° in the
    circuit of Figure 28–1, the power dissipated by the load resistor   (   )
    (a) decreases   (b) increases   (c) remains the same

29

THE UNIJUNCTION TRANSISTOR

PURPOSE AND BACKGROUND

The purposes of this experiment are (1) to demonstrate how to test a unijunction transistor (UJT) with an ohmmeter and (2) to describe the operation of a relaxation oscillator. The UJT is a three-terminal switching device used primarily as an oscillator. Although it uses the name *transistor* and its schematic symbol closely resembles that of an n-channel FET, the UJT, unlike other transistors, cannot be used as an amplifier.

Text Reference: 11–6, The Unijunction Transistor (UJT).

REQUIRED PARTS AND EQUIPMENT

Resistors (1/4 W):
- [ ] 68 kΩ
- [ ] 1 kΩ
- [ ] 47 kΩ
- [ ] 0.01-μF capacitor
- [ ] 2N2646 unijunction transistor (HEP 310, MU020, or equivalent)

- [ ] 0–15 V dc power supply
- [ ] VOM or DMM (preferred)
- [ ] Dual trace oscilloscope
- [ ] Breadboarding socket

USEFUL FORMULAS

Approximate output frequency

$$(1) \quad f_o \simeq \frac{1}{RC \ln\left(\dfrac{1}{1-\eta}\right)} \quad \text{(Hz)}$$

where η = intrinsic standoff ratio

Peak capacitor voltage

$$(2) \quad V_p = \eta V_{BB} + V_{pn}$$

where V_{BB} = UJT interbase voltage $(V_{B2} - V_{B1})$

V_{pn} = UJT emitter-base diode voltage

Base 2 dc voltage

$$(3) \quad V_{B2} = \frac{r_{BB} V_{CC}}{r_{BB} + R_1 + R_2}$$

Base 1 dc voltage

$$(4) \quad V_{B1} = \frac{R_1 V_{CC}}{r_{BB} + R_1 + R_2}$$

PROCEDURE

1. As with diodes and transistors (see Experiments 1 and 8), VOMs can be used to check quickly whether a UJT is good or bad. Unless they have a specific function for this purpose, *DMMs are usually not able to perform this test properly.* Refer to Experiment 8 to determine the polarity of your ohmmeter's leads.

2. Using a 2N2646 UJT, whose pin diagram and diode junction representation are shown in Figure 29–1, connect the ohmmeter's positive lead to the transistor's base 1 (*B*1) lead while the ohmmeter's negative lead is connected to the base 2 (*B*2) lead. Measure the resistance and record this value in Table 29–1.

3. Reverse the meter's leads so that the positive lead is connected to the *B*2 lead while the ohmmeter's negative lead is connected to the *B*1 lead. Measure the resistance and record this value in Table 29–1.

   The meter readings in both cases should be the same. This resistance, called the *base-to-base resistance* r_{BB}, is typically 4 kΩ to 10 kΩ.

4. Now connect the meter's positive lead to the emitter lead (*E*), and the negative lead to the *B*1 lead. Measure the resistance and record this value in Table 29–1.

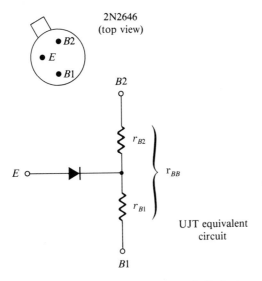

FIGURE 29–1 *Pin diagram and diode junction representation of 2N2646 UJT.*

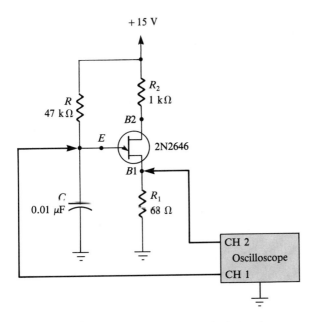

FIGURE 29–2 *Schematic diagram of circuit.*

5. Reverse the meter leads. Measure the resistance and record this value in Table 29–1.

6. Again connect the meter's positive lead to the emitter lead, but now connect the meter's negative lead to *B*2. Measure the resistance and record this value in Table 29–1.

263

7. Reverse the meter leads. Measure the resistance and record this value in Table 29–1.

8. In Steps 4 and 6, you separately forward biased both emitter-base junctions. In Steps 5 and 7, these junctions were reverse biased. Consequently, the resistance reading obtained in Step 4 should be less than in Step 5, and the reading in Step 6 should be less than in Step 7. In general, one of the emitter-base resistance readings will be lower than the other. Usually the emitter-$B2$ junction exhibits the lower resistance.

   When you are not sure which leads of a UJT are the emitter, $B1$, and $B2$ leads, you should be able to identify the two base leads quickly since they measure the same resistance with either meter polarity (Steps 2 and 3).

9. Wire the circuit shown in the schematic diagram of Figure 29–2. Initially, leave the 47-kΩ timing resistor disconnected from the circuit. In addition, set your oscilloscope to the following approximate settings:

   Channel 1: 5 V/division, dc coupling
   Channel 2: 2 V/division, dc coupling
   Time base: 0.2 ms/division

10. Apply power to the breadboard. First measure the dc voltages at $B1$ and $B2$ with respect to ground, and then take the difference, $V_{B2} - V_{B1}$, to obtain the *interbase voltage* V_{BB}. Do this for both the measured and the expected values, and record them in Table 29–2. Using the experimental value you measured in Steps 2 and 3 for the *interbase resistance* r_{BB} and Equations 3 and 4 given in the "Useful Formulas" section, compare the values for V_{B1} and V_{B2} with their expected values.

11. Now add the 47-kΩ timing resistor to the circuit as shown in Figure 29–2. On Channel 1, you should now see displayed a capacitor charge-discharge "sawtooth" waveform like that shown in Figure 29–3. Using your oscilloscope, measure the frequency (f_o) of this waveform, and record this value in Table 29–2.

12. Using the capacitor voltage waveform shown in Figure 29–4 as a guide, measure the maximum voltage (V_p) and the minimum voltage (V_v) with respect to ground, and record your results in Table 29–2. Using the experimental value for V_p and assuming a diode voltage V_D of 0.7 V, determine the UJT's intrinsic standoff ratio (η) from Equation 2. Record this value in Table 29–2.

13. Using the value for the intrinsic standoff ratio found in Step 12, as well as timing resistor R and capacitor C, calculate the expected output frequency of the UJT relaxation oscillator (Equation 1), recording this value in Table 29–2. Compare this value with that measured in Step 11.

14. Now notice the positive-going waveform on Channel 2, which is the voltage across resistor R_1. Using the waveform shown in

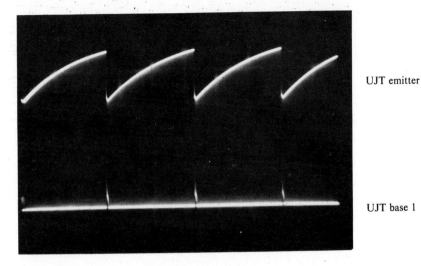

UJT emitter

UJT base 1

FIGURE 29–3

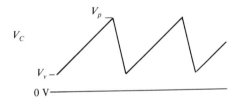

FIGURE 29–4   *Capacitor voltage waveform.*

Figure 29–5 as a guide, measure the maximum voltage (V_1) and the minimum voltage (V_2) with respect to ground, and record your values in Table 29–2. You should find that the minimum voltage is approximately equal to the base 1 dc voltage measured in Table 29–1.

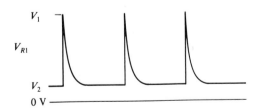

FIGURE 29–5   *UJT base 1 voltage waveform.*

15. Now connect the Channel 2 probe to $B2$ of the UJT. You should observe a negative-going waveform on Channel 2, which is the voltage across resistor R_2. Using the waveform shown in Figure 29–6 as a guide, measure the maximum voltage (V_3) and the

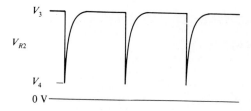

FIGURE 29–6 *UJT base 2 voltage waveform.*

minimum voltage (V_4) with respect to ground, and record your values in Table 29–2. You should find that the maximum voltage is approximately equal to the base 2 dc voltage measured in Table 29–1, while the minimum voltage is approximately equal to the *valley voltage*, V_v.

Consequently, we are able to obtain three separate waveforms from the UJT relaxation oscillator.

WHAT YOU HAVE DONE

This experiment demonstrated how to properly test a unijunction transistor (UJT) with an ohmmeter. In addition, this experiment demonstrated the operation of a relaxation oscillator, identifying the waveforms present at all three terminals of the UJT. In addition, the intrinsic standoff ratio (η) of a UJT was determined.

Name _____ Date _____

THE UNIJUNCTION TRANSISTOR

OBJECTIVES/PURPOSE:

SCHEMATIC DIAGRAM:

Name _____   Date _____

DATA FOR EXPERMIMENT 29

TABLE 29–1 *UJT resistance tests.*

Step Number	Ohmmeter Leads +	Ohmmeter Leads −	Resistance
2	base 1	base 2	
3	base 2	base 1	
4	emitter	base 1	
5	base 1	emitter	
6	emitter	base 2	
7	base 2	emitter	

Name _____   Date _____

TABLE 29–2 *UJT relaxation oscillator.*

Parameter	Measured Value	Expected Value	% Error
V_{B2}			
V_{B1}			
V_{BB}			
V_p			
V_v			
f_o			
η			
V_1			
V_2			
V_3			
V_4			

Name _____ Date _____

RESULTS AND CONCLUSIONS:

REVIEW QUESTIONS FOR EXPERIMENT 29

1. If the intrinsic standoff ratio for the UJT used in the circuit of Figure 29–2 is 0.75, the output frequency of the relaxation oscillator is approximately
   (a) 670 Hz   (b) 1120 Hz   (c) 1530 Hz   (d) 2127 Hz   (   )
2. If the intrinsic standoff ratio for the UJT used in the circuit of Figure 29–2 increases, the output frequency of the relaxation oscillator
   (a) decreases   (b) increases
   (c) remains essentially the same   (   )
3. If the UJT interbase resistance is 6 kΩ, the base 2 voltage for the circuit the Figure 29–2 is approximately
   (a) 0.15 V   (b) 2 V   (c) 13 V   (d) 14.85 V   (   )
4. The waveform at the base 2 lead of a UJT oscillator is a
   (a) sawtooth
   (b) positive-going spike
   (c) negative-going spike
   (d) square wave   (   )
5. The waveform at the emitter lead of a UJT oscillator is a
   (a) sawtooth
   (b) positive-going spike
   (c) negative-going spike
   (d) square wave   (   )

30

OP-AMP SLEW RATE

PURPOSE AND BACKGROUND

The purpose of this experiment is to measure the slew rate of a 741 operational amplifier. The slew rate, a *large-signal* parameter, is defined as the maximum time rate of change of the output voltage of an op-amp in response to a step input voltage. Expressed in units of volts per microsecond (V/μs), the slew rate is dependent upon the frequency response of the internal stages of the op-amp. Thus, the higher the slew rate, the better the frequency response of the amplifier.

The measurement of the operational amplifier's slew rate is always accomplished with a large-signal amplifier having *unity gain* with a high input frequency signal.

Text Reference: 12–3, Op-Amp Parameters.

REQUIRED PARTS AND EQUIPMENT

- ☐ Two 10-kΩ resistors, 1/4 W
- ☐ 741 op-amp (8-pin mini-DIP)
- ☐ Two 0–15 V dc power supplies
- ☐ Signal generator
- ☐ Dual trace oscilloscope
- ☐ Breadboarding socket

PROCEDURE

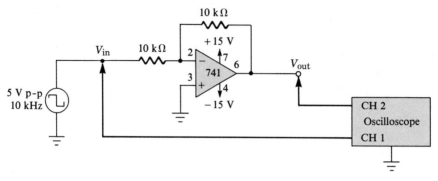

FIGURE 30–1 *Schematic diagram of circuit.*

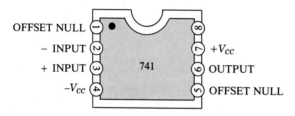

FIGURE 30–2 *Pin diagram of 741 op-amp.*

1. Wire the circuit shown in the schematic diagram of Figure 30–1, and set your oscilloscope for the following approximate settings:

   Channel 1: 5 V/division, ac coupling
   Channel 2: 1 V/division, ac coupling
   Time base: 10 μs/division

2. Apply power to the breadboard, and adjust the square-wave input signal at 5 V peak-to-peak with a frequency of 10 kHz. The output signal should have a trapezoidal shape, as shown in Figure 30–3. If the operational amplifier were ideal (that is, infinite slew rate), the output signal would look exactly the same as the input at very high frequencies. In reality, it takes a finite amount of time for the large-signal amplifier to switch from one voltage extreme to the other.

3. Measure the peak-to-peak output voltage, ΔV, and record your result in Table 30–1.

4. Measure Δt, the time in microseconds that it takes the output voltage to swing from its minimum to its maximum value, or vice versa, and record this value in Table 30–1.

5. From the measurements in Steps 3 and 4, calculate the slew rate, $\Delta V / \Delta t$, for your 741 amplifier, and record your result in Table 30–1.

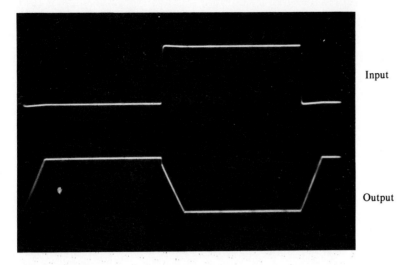

Input

Output

FIGURE 30–3

For the 741 operational amplifier, the typical slew rate is 0.5 V/μs. Other op-amps, such as the LM318, are faster, having a typical slew rate of 70 V/μs. This is 140 times better than that of the 741 op-amp.

WHAT YOU HAVE DONE

This experiment demonstrated how to measure the slew rate, or time rate of change of the output voltage of a 741 operational amplifier. It is a large-signal parameter and is measured using a closed-loop gain of 1.

NOTES

Name ______________________________ Date ____________

OP-AMP SLEW RATE

OBJECTIVES/PURPOSE:

SCHEMATIC DIAGRAM:

DATA FOR EXPERIMENT 30

TABLE 30–1

ΔV	V
Δt	μs
Slew Rate	V/μs

RESULTS AND CONCLUSIONS:

REVIEW QUESTIONS FOR EXPERIMENT 30

1. For the circuit of Figure 30–1, using a ± 15 V supply, the maximum possible output voltage swing is approximately
    (a) 5 V   **(b)** 15 V   **(c)** 20 V   **(d)** 30 V   (   )
2. The maximum time rate of change of the output voltage of the circuit of Figure 30–1 in response to a step input is termed the
    (a) gain-bandwidth product
    (b) slew rate
    (c) output voltage swing
    (d) common-mode rejection ratio   (   )
3. The slew rate is usually specified in units of
    (a) V/s   **(b)** V/μs   **(c)** dB   **(d)** MHz   (   )
4. For an operational amplifier, the slew rate limits the
    (a) input impedance   **(b)** common-mode rejection
    (c) voltage gain   **(d)** frequency response   (   )
5. For the circuit of Figure 30–1, if the output voltage swings from +5V to -10V in 0.5 μs, the slew rate is
    (a) 5 V/μs   **(b)** 15 V/μs   **(c)** 20 V/μs   **(d)** 30 V/μs   (   )

NOTES

31

OP-AMP COMMON-MODE REJECTION

PURPOSE AND BACKGROUND

The purpose of this experiment is to measure the common-mode rejection of a 741 operational amplifier. If the same signal is applied simultaneously to both inputs, called the *common-mode input,* then the output voltage of an ideal op-amp should be zero. Since operational amplifiers are not ideal devices, a small but finite output voltage will be present when both the input voltages are the same. The ratio of the common-mode input voltage to the generated output voltage is termed the *common-mode rejection,* or CMR, and is expressed in decibels. The higher the CMR, the better the rejection and the smaller the output voltage.

Text References: 12–2, The Differential Amplifier; 12–3, Op-Amp Parameters.

REQUIRED PARTS AND EQUIPMENT

Resistors (1/4 W):
- ☐ Two 100 Ω
- ☐ 10 kΩ
- ☐ Two 100 kΩ
- ☐ 100-kΩ potentiometer, or 10-turn trim-pot

- ☐ 741 op-amp (8-pin mini-DIP)
- ☐ Two 0–15 V dc power supplies
- ☐ VOM or DMM (preferred)
- ☐ Signal generator
- ☐ Dual trace oscilloscope
- ☐ Breadboarding socket

279

USEFUL FORMULAS

Differential amplifier voltage gain

$$\text{(1)} \quad A_d = \frac{R_2}{R_1} \qquad \text{(where } R_1 = R_3 \text{ and } R_2 = R_4)$$

Common-mode voltage gain

$$\text{(2)} \quad A_{cm} = \frac{V_{out(cm)}}{V_{in(cm)}}$$

dB common-mode rejection

$$\text{(3)} \quad \text{CMR (dB)} = 20 \log \left(\frac{A_{v(d)}}{A_{cm}} \right)$$

PROCEDURE

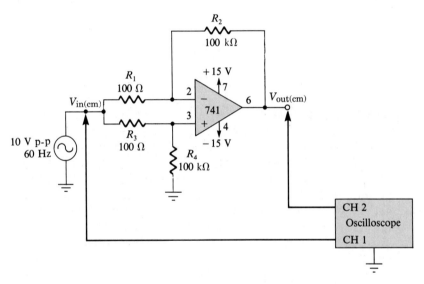

FIGURE 31–1 *Schematic diagram of circuit.*

1. Wire the circuit shown in the schematic diagram of Figure 31–1, and set your oscilloscope for the following approximate settings:

    Channel 1: 2 V/division, ac coupling
    Channel 2: 0.02 V/division, ac coupling
    Time base: 5 ms/division

2. Apply power to the breadboard. Adjust the input voltage, called the *common-mode input voltage, V*$_{in(cm)}$, to 10 V peak-to-peak at a frequency of approximately 60 Hz. You should make this voltage setting as accurate as possible.

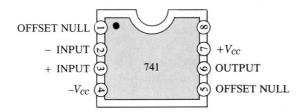

FIGURE 31–2 *Pin diagram of 741 op-amp.*

3. Using a VOM or DMM, now measure the rms common-mode input and output voltages, recording your results in Table 31–1. From the common-mode input and output voltages, calculate the common-mode voltage gain, A_{cm} (Equation 2), recording this result in Table 31–1.

4. This circuit is called a *difference,* or *differential amplifier,* and the differential voltage gain, $A_{v(d)}$, is based upon all four resistors, as given in the "Useful Formulas" section of this experiment (Equation 1). First calculate the differential voltage gain, and then use it to calculate the common-mode rejection (in decibels) for your 741 operational amplifier. Record your result in Table 31–1.

    Most manufacturers of the 741 op-amp cite a minimum CMR of 70 dB, although a value of 90 dB is typical.

5. In some instances, the CMR may be improved by trimming one or more resistors of the circuit of Figure 31–1. Disconnect the signal generator and dc power to the circuit. Replace R_4 with a series connection of a 100-kΩ potentiometer and a 10-kΩ resistor.

6. Apply power to the breadboard and again adjust the common-mode input voltage to 10 V peak-to-peak at a frequency of 60 Hz.

7. Using the oscilloscope to observe the output of the operational amplifier at pin 6, adjust the 100-kΩ potentiometer for a minimum output voltage.

8. Repeat Steps 3 and 4 using a differential gain of 1000. Record your results in Table 31–2. Do you see an improvement in the CMR?

WHAT YOU HAVE DONE

This experiment demonstrated how to measure the common-mode rejection (CMR) of a 741 operational amplifier and how to maximize the CMR of a difference amplifier. For an ideal operational amplifier, the common-mode output voltage should be zero, giving an infinite value for the CMR. The ratio of the common-mode input voltage to the generated output voltage is termed the *common-mode rejection ratio,* or *CMRR.* When expressed in decibels, it is called the *common-mode rejection (CMR).* The higher the CMR, the better the rejection, resulting in a smaller output voltage.

Name _____   Date _____

OP-AMP COMMON-MODE REJECTION

OBJECTIVES/PURPOSE:

SCHEMATIC DIAGRAM:

DATA FOR EXPERIMENT 31

TABLE 31–1

Measured common-mode input voltage, $V_{in(cm)}$	V
Measured common-mode output voltage, $V_{out(cm)}$	V
Calculated common-mode voltage gain, A_{cm}	
Calculated differential voltage gain, $A_{v(d)}$	
Calculated common-mode rejection, CMR	dB

TABLE 31–2

Measured common-mode input voltage, $V_{in(cm)}$	V
Measured common-mode output voltage, $V_{out(cm)}$	V
Calculated common-mode voltage gain, A_{cm}	
Calculated differential voltage gain, $A_{v(d)}$	
Calculated common-mode rejection, CMR	dB

Name _____ Date _____

RESULTS AND CONCLUSIONS:

REVIEW QUESTIONS FOR EXPERIMENT 31

1. In a differential amplifier, the signal applied simultaneously to
    both inputs is the
    (a) differential input   **(b)** noninverting input
    (c) inverting input   **(d)** common-mode input   ( )
2. An increase in common-mode rejection means an increase in
    the amplifier's
    (a) input impedance   **(b)** frequency response
    (c) voltage gain   **(d)** noise immunity   ( )
3. Differential-amplifier common-mode rejection is measured in
    (a) V   **(b)** V/μs   **(c)** dB   **(d)** V/mV   ( )
4. If the differential voltage gain is 100 and the common-mode
    voltage gain is 0.001, the common-mode rejection is
    (a) 40 db   **(b)** 60 db   **(c)** 80 db   **(d)** 100 db   ( )
5. For the circuit of Figure 31–1, if the common-mode rejection
    ratio is 100,000 : 1 and the input voltage is 10 V peak-to-peak,
    the peak-to-peak output voltage is
    (a) 0.001 V   **(b)** 0.01 V   **(c)** 0.1 V   **(d)** 1 V   ( )

NOTES

32

OP-AMP INVERTING AND NONINVERTING AMPLIFIERS

PURPOSE AND BACKGROUND

The purpose of this experiment is to demonstrate the operation of both inverting and noninverting amplifier circuits using a 741 operational amplifier. Both circuits operate in the closed-loop mode. The inverting amplifier's closed-loop voltage gain can be less than, equal to, or greater than 1. As its name implies, its output signal is always inverted with respect to its input signal. On the other hand, the non-inverting amplifier's closed-loop voltage gain is always greater than 1, while the input and output signals are always in-phase.

Text Reference: 12–5, Op-Amp Configurations with Negative Feedback.

REQUIRED PARTS AND EQUIPMENT

Resistors (1/4 W):
- [ ] 1 kΩ
- [ ] 4.7 kΩ
- [ ] Two 10 kΩ
- [ ] 22 kΩ
- [ ] 47 kΩ
- [ ] 100 kΩ

- [ ] 741 op-amp (8-pin mini-DIP)
- [ ] Two 0–15 V dc power supplies
- [ ] Signal generator
- [ ] Dual trace oscilloscope
- [ ] Breadboarding socket

287

USEFUL FORMULAS

Inverting amplifier closed-loop voltage gain

(1) $A_v = -\dfrac{R_f}{R_i}$

Noninverting amplifier closed-loop voltage gain

(2) $A_v = 1 + \dfrac{R_f}{R_i}$

PROCEDURE

1. Wire the inverting amplifier circuit shown in the schematic diagram of Figure 32–1A, and set your oscilloscope for the following approximate settings:

   Channels 1 and 2: 0.5 V/division, ac coupling
   Time base: 1 ms/division

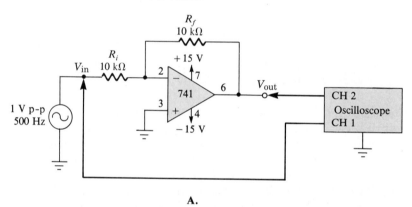

A.

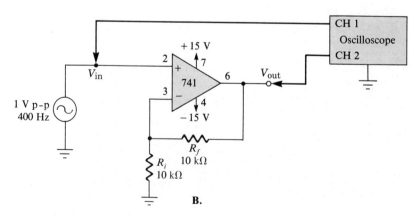

B.

FIGURE 32–1 *Schematic diagram of circuits.*

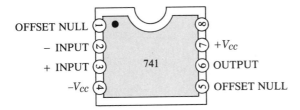

FIGURE 32-2 *Pin diagram of 741 op-amp.*

2. Apply power to the breadboard, and adjust the input voltage to 1 V peak-to-peak and the frequency at 500 Hz. Position the input voltage above the output voltage on the oscilloscope's display. What is the difference between the two signals?

   Notice that the output signal is of opposite form, or is *inverted,* compared to the input signal, as shown in Figure 32–3. The output voltage is then said to be inverted from, or 180° out-of-phase with, the input, since the positive peak of the output signal occurs when the input's peak is negative.

3. Measure the peak-to-peak output voltage. Then determine the voltage gain and compare it to the expected value, recording your results in Table 32–1.

   The peak-to-peak output voltage should be the same as the input (1 V), so that the voltage gain is −1. The minus sign indicates that the output is *inverted* with respect to the input.

4. Keeping the input signal at 1 V peak-to-peak, change resistor R_f according to Table 32–1, recording your results as in Step 3. Each time, disconnect the power supplies and signal generator before you change the resistor. Do your results agree with those obtained from the equation for the inverting amplifier voltage gain (Equation 1)?

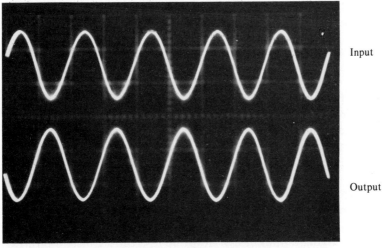

Input

Output

FIGURE 32-3

As the results of Table 32–1 indicate, the voltage gain of an inverting amplifier can be made to be less than 1, equal to 1, or greater than 1.

5. Now wire the noninverting amplifier circuit shown in the schematic diagram of Figure 32–1B. Apply power to the breadboard and adjust the input voltage to 1 V peak-to-peak and the frequency at 400 Hz. Again position the input voltage above the output voltage on the oscilloscope's display. What is the difference between the two signals?

    The only difference is that the output signal is *larger* than the input signal, as shown in Figure 32–4. Both signals are said to be *in-phase*, since the output signal goes positive exactly when the input does.

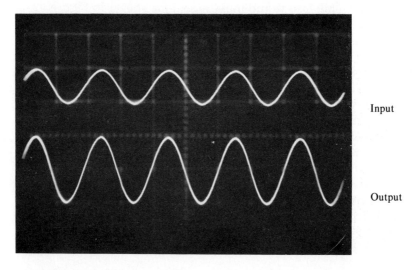

Input

Output

FIGURE 32–4

6. Measure the peak-to-peak output voltage. Then determine the voltage gain and compare it to the expected value, recording your results in Table 32–2.

    The peak-to-peak output voltage should be approximately 2 V, so that the voltage gain is 2.

7. Keeping the input signal at 1 V peak-to-peak, change resistor R_f according to Table 32–2, recording your results as in Step 6. Each time, disconnect the power supplies and signal generator before you change the resistor. Do your results agree with those obtained from the equation for the noninverting amplifier voltage gain (Equation 2)?

    As the results of Table 32–2 indicate, the voltage gain of a noninverting amplifier can never be less than 1 or equal to 1. It will always be greater than 1.

WHAT YOU HAVE DONE

This experiment demonstrated and compared the operation of the noninverting and inverting amplifier circuits using the 741 operational amplifier. Both circuits are termed *linear amplifiers,* as the output signal is linearly proportional to its input signal. The inverting amplifier's closed-loop voltage gain can be less than, equal to, or greater than 1 and its output signal is always of the opposite polarity, or inverted with respect to its input signal giving a phase shift of 180°. On the other hand, the noninverting amplifier's closed-loop voltage gain is always greater than 1, while the input and output signals are always in phase.

NOTES

Name _____   Date _____

OP-AMP INVERTING AND NONINVERTING AMPLIFIERS

OBJECTIVES/PURPOSE:

SCHEMATIC DIAGRAM:

Name _____ Date _____

DATA FOR EXPERIMENT 32

TABLE 32–1 *Inverting amplifier.*

R_f	Measured V_{out}	Measured Gain	Expected Gain	% Error
10 kΩ				
22 kΩ				
47 kΩ				
100 kΩ				
4.7 kΩ				
1 kΩ				

TABLE 32–2 *Noninverting amplifier.*

R_f	Measured V_{out}	Measured Gain	Expected Gain	% Error
10 kΩ				
22 kΩ				
47 kΩ				
100 kΩ				
4.7 kΩ				
1 kΩ				

Name _____ Date _____

RESULTS AND CONCLUSIONS:

REVIEW QUESTIONS FOR EXPERIMENT 32

1. The circuit of Figure 32–1A is
   (a) an inverting amplifier   **(b)** a noninverting amplifier
   (c) a differential amplifier   **(d)** a voltage follower   (   )
2. The voltage gain of an inverting amplifier can be
   (a) less than 1   **(b)** equal to 1
   (c) greater than 1   **(d)** all of the above   (   )
3. The output signal of an inverting amplifier is out-of-phase with
   its input signal by
   (a) 0°   **(b)** 90°   **(c)** 180°   **(d)** 270°   (   )
4. The voltage gain of a noninverting amplifier can be
   (a) less than 1   **(b)** equal to 1
   (c) greater than 1   **(d)** all of the above   (   )
5. The output signal of a noninverting amplifier is out-of-phase
   with its input signal by
   (a) 0°   **(b)** 90°   **(c)** 180°   **(d)** 270°   (   )

     295

NOTES

33

OP-AMP COMPARATORS

PURPOSE AND BACKGROUND

The purpose of this experiment is to demonstrate the operation of noninverting and inverting comparator circuits using a 741 operational amplifier. A comparator determines whether an input voltage is greater than a predetermined reference level. Since a comparator operates in an open-loop mode, the output voltage approaches either its positive or its negative supply voltage.

Text Reference: 14–1, Comparators.

REQUIRED PARTS AND EQUIPMENT

Resistors (1/4 W):
- ☐ Two 1 kΩ
- ☐ 4.7 kΩ
- ☐ Two 10 kΩ
- ☐ 47 kΩ
- ☐ 100 kΩ
- ☐ 100-kΩ potentiometer
- ☐ 741 op-amp (8-pin mini-DIP)

- ☐ 1N914 (or 1N4148) diode
- ☐ LED
- ☐ 2N3904 npn transistor
- ☐ Two 0–15 V dc power supplies
- ☐ Signal generator
- ☐ Dual trace oscilloscope
- ☐ Breadboarding socket

USEFUL FORMULAS

Noninverting comparator output

    (1) $V_{out} = +V_{SAT}$ when $V_{in} > V_{REF}$

    (2) $V_{out} = -V_{SAT}$ when $V_{in} < V_{REF}$

Inverting comparator output

    (3) $V_{out} = +V_{SAT}$ when $V_{in} < V_{REF}$

    (4) $V_{out} = -V_{SAT}$ when $V_{in} > V_{REF}$

PROCEDURE

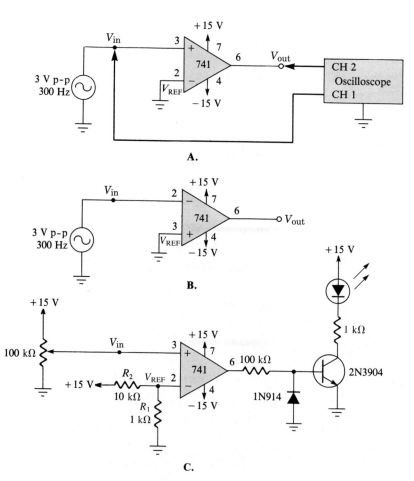

FIGURE 33–1 *Schematic diagram of circuits.*

1. Wire the circuit shown in the schematic diagram of Figure 33–1A, and set your oscilloscope for the following approximate settings:

> Channel 1: 1 V/division, dc coupling
> Channel 2: 10 V/division, dc coupling
> Time base: 1 ms/division

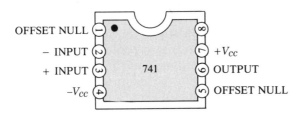

FIGURE 33–2 *Pin diagram of 741 op-amp.*

2. Apply power to the breadboard, and adjust the input voltage at 3 V peak-to-peak and the frequency at 300 Hz. What is the polarity of the output voltage when the input signal goes positive? When the input goes negative?

When the input signal V_{in} is applied to the op-amp's noninverting input, the output signal's polarity will be the *same* as that of the input, so that this circuit is a *noninverting comparator*. In this case, the reference voltage V_{REF} is zero (the inverting input is grounded). Because of the high open-loop gain of the

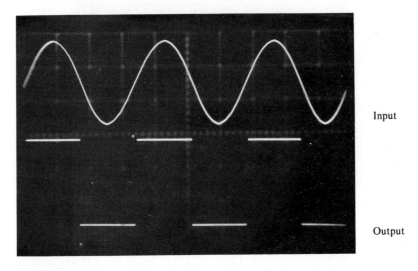

Input

Output

FIGURE 33–3

op-amp, the output immediately goes positive when V_{in} is greater than zero volts (V_{REF}), and vice versa, as shown in Figure 33–3. This circuit is also referred to as a *noninverting zero-level detector* since it detects the polarity of the input signal.

The maximum output (saturation) voltage, V_{SAT}, for the 741 op-amp is typically 12 V to 14 V when using a 15-V supply.

3. Disconnect the power and signal leads to the breadboard, and reverse the input connections to the op-amp so that the input signal is now connected to the inverting input while the noninverting input is grounded, as shown in Figure 33–1B.

4. Again apply both the power and signal leads to the breadboard. Now what is the difference between the operation of this circuit and that of the circuit used earlier?

Notice that the output of this comparator circuit has a polarity that is *inverted* with respect to the input signal. Such a circuit is called an *inverting comparator*. Furthermore, since the reference voltage (the voltage at the noninverting input) is zero, this circuit is also referred to as an *inverting zero-level detector*. When the polarity of the input signal is positive, the output voltage equals $-V_{SAT}$, and vice versa, as shown in Figure 33–4. As can be seen, both circuits are useful in converting sine waves into square waves.

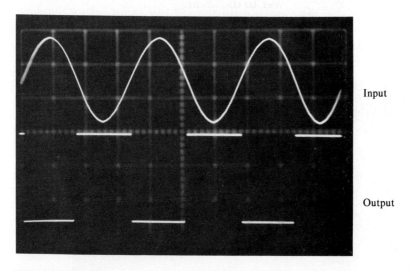

Input

Output

FIGURE 33–4

5. Disconnect the power and signal leads from the breadboard, and wire the circuit shown in Figure 33–1C. Make sure that you have the 1N914 diode and LED, as well as the npn transistor, wired correctly.

6. Apply power to the breadboard. Depending on the setting of the potentiometer, the LED may or may not be lit when you connect the power. If the LED is *on*, turn the potentiometer past the point at which the LED is off.
7. With your oscilloscope, measure the voltage at the op-amp's inverting terminal (pin 2), which is the reference voltage V_{REF} for the comparator, and record the value in Table 33–1.
8. Now connect the oscilloscope to the op-amp's noninverting input (pin 2). While watching the LED, vary the potentiometer just until the LED lights up. Measure this voltage at pin 3, $V_{in(on)}$, and record your result in Table 33–1. How does this value compare with the one you determined in Step 7?

    These two values should be nearly the same. When the input voltage V_{in} at the noninverting input exceeds the comparator's reference voltage at the inverting input, the op-amp comparator's output switches from its negative saturation voltage to its positive saturation voltage. This circuit is a *noninverting comparator* whose nonzero reference voltage is set by the 10-kΩ and 1-kΩ resistors connected as a simple voltage divider. The transistor-LED circuit connected to the output of the comparator allows you to determine visually whether the input voltage is greater or less than the reference voltage. If the input voltage exceeds the reference, the LED is lit.
9. Disconnect the power to the breadboard. Verify the operation of this noninverting comparator by varying voltage-divider resistor R_1 and repeating Steps 6, 7, and 8, according to Table 33–2.

WHAT YOU HAVE DONE

This experiment demonstrated the operation of noninverting and inverting comparator circuits using a 741 operational amplifier. It was shown that the basic comparator determines if an input voltage is greater than a predetermined reference level. Since comparators detect whether or not its input signal either exceeds or drops below a given voltage level, they are also referred to as *level detectors*. Comparators operate in the open-loop mode, and as a consequence, the output voltage always approaches either its positive or negative supply voltage level.

NOTES

Name ______________________________ Date ____________

OP-AMP COMPARATORS

OBJECTIVES/PURPOSE:

SCHEMATIC DIAGRAM:

Name _____   Date _____

DATA FOR EXPERIMENT 33

TABLE 33–1 *Inverting comparator.*

R_1	Measured V_{REF}	Measured $V_{in(on)}$
1 kΩ		
10 kΩ		
47 kΩ		

TABLE 33–2 *Noninverting comparator.*

R_1	Measured V_{REF}	Measured $V_{in(on)}$
1 kΩ		
10 kΩ		
47 kΩ		

Name _____   Date _____

RESULTS AND CONCLUSIONS:

REVIEW QUESTIONS FOR EXPERIMENT 33

1. The circuit of Figure 33–1A is
   (a) an inverting comparator
   (b) a noninverting comparator   (   )
2. The reference voltage for the comparator of Figure 33–1A is
   (a) 0 V   (b) −15 V
   (c) +15 V   (d) none of the above   (   )
3. For the circuit of Figure 33–1B, if the input voltage is greater than the reference voltage, the output voltage is approximately
   (a) −13 V   (b) −3 V   (c) +3 V   (d) +13 V   (   )
4. For the circuit of Figure 33–1B, if the input signal is a sine wave, the output signal looks like a
   (a) sine wave
   (b) sine wave, but inverted with respect to the input
   (c) square wave
   (d) square wave, but inverted with respect to the input   (   )
5. For the circuit of Figure 33–1C, if R_1 and R_2 are 10 kΩ, the LED is lit when the input voltage is
   (a) less than −7.5 V
   (b) 0 V
   (c) greater than 7.5 V
   (d) any voltage between −7.5 V and +7.5 V   (   )

NOTES

34

OP-AMP DIFFERENTIATOR AND INTEGRATOR

PURPOSE AND BACKGROUND

The purpose of this experiment is to demonstrate the operation of both a differentiator and an integrator using an op-amp. A differentiator is a circuit that calculates the instantaneous slope of the line at every point on a waveform. On the other hand, an integrator computes the area underneath the curve of a given waveform. Differentiation and integration are *paired* mathematical operations in that one has the opposite effect of the other. For example, if you integrate a waveform and then differentiate it, you obtain the original waveform.

Text Reference: 14–3, The Integrator and Differentiator.

REQUIRED PARTS AND EQUIPMENT

Resistors (1/4 W):
- [ ] 2.2 kΩ
- [ ] Two 10 kΩ
- [ ] 22 kΩ
- [ ] 100 kΩ

Capacitors:
- [ ] 0.0022 μF
- [ ] 0.0047 μF

- [ ] 741 op-amp (8-pin mini-DIP)
- [ ] Two 0–15 V dc power supplies
- [ ] Function generator
- [ ] Dual trace oscilloscope
- [ ] Breadboarding socket

USEFUL FORMULAS

Differentiator

Output voltage:

(1) $V_{out} = -R_F C \left(\dfrac{dV_{in}}{dt} \right)$

Low-frequency response:

(2) $f_c \doteq \dfrac{1}{2\pi R_S C}$

When $f_{in} < f_c$, the circuit acts as a differentiator.
When $f_{in} > f_c$, the circuit approaches an inverting amplifier
with a voltage gain of $-R_F/R_S$.

Integrator

Output voltage:

(3) $V_{out} = -\dfrac{1}{R_1 C} \displaystyle\int_0^t V_{in} dt$

Low-frequency response:

(4) $f_c = \dfrac{1}{2\pi R_S C}$

When $f_{in} > f_c$, the circuit acts as an integrator.
When $f_{in} < f_c$, the circuit approaches an inverting amplifier
with a voltage gain of $-R_S/R_1$.

For minimum output offset due to input bias currents:

(5) $R_2 = \dfrac{R_1 R_S}{R_1 + R_S}$

PROCEDURE

1. Wire the differentiator circuit shown in the schematic diagram
   of Figure 34–1A, and set your oscilloscope to the following approximate settings:

   > Channel 1: 0.5 V/division, dc coupling
   > Channel 2: 0.05 V/division, dc coupling
   > Time base: 0.5 ms/division

2. Apply power to the breadboard, and adjust the peak-to-peak
   voltage of the input triangle wave at 1 V and the frequency at
   400 Hz. As shown in Figure 34–3, the output signal is a square
   wave that is 180° out-of-phase with the input signal.

3. Temporarily remove the probe connected to Channel 2 of the oscilloscope, and adjust the resulting straight line (ground level)

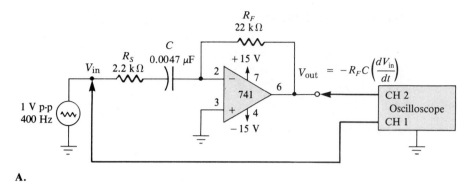

A.

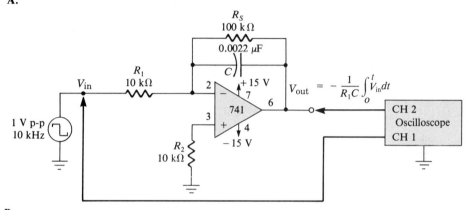

B.

FIGURE 34–1 *Schematic diagram of circuits.*

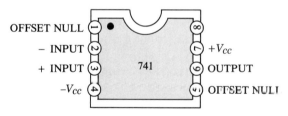

FIGURE 34–2 *Pin diagram of 741 op-amp.*

at some convenient position on the screen. Reconnect the probe to the output of the differentiator, and measure the negative peak voltage (with respect to ground) of the square wave, recording your result in Table 34–1.

4. Now measure the time duration for which the square-wave signal is negative (t_1). The peak voltage of a square wave that results from differentiating a triangle waveform having a peak voltage V_m is given by

$$V_{out}(\text{peak}) = -\frac{2R_F C V_M}{t_1}$$

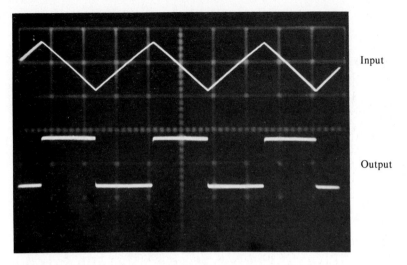

Input

Output

FIGURE 34–3

Compute the expected value of the negative peak voltage, and compare it with the measured value above. Record your results in Table 34–1.

5. Change the time base to 0.2 ms/division and Channel 2 to 0.1 V/division. Then adjust the input frequency at 1 kHz. Repeat Steps 3 and 4. You should find that the peak output voltage increases.

6. Now change the input frequency to 30 kHz. Adjust the time base to 10 μs/division and Channel 2 to 2 V/division. What does the output signal look like?

   Notice that the output signal looks like a triangle wave with a phase shift of 180°. Why?

   Above approximately 15.4 kHz, the circuit ceases to act as a differentiator since the reactance of the 0.0047-μF capacitor is now less than that of the 2.2-kΩ resistor (R_S). Above this frequency, the circuit functions like that of an inverting amplifier having a voltage gain of $-R_F/R_S$.

7. Measure the peak-to-peak output voltage and determine the voltage gain, recording your values in Table 34–1. How does the voltage gain compare to that of an inverting amplifier?

8. Wire the integrator circuit shown in the schematic diagram of Figure 34–1B, and set your oscilloscope to the following approximate settings:

   Channels 1 and 2: 0.5 V/division, dc coupling
   Time base: 20 μs/division

9. Apply power to the breadboard, and adjust the peak-to-peak voltage of the input square wave at 1 V and the frequency

at 10 kHz. As shown in Figure 34–4, the output signal is a triangle wave that is 180° out-of-phase with the input signal.

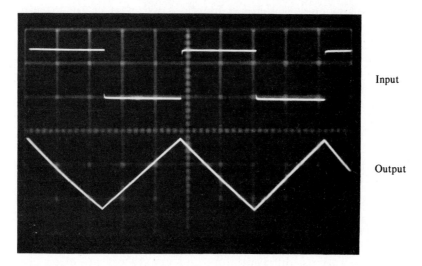

Input

Output

FIGURE 34–4

10. Temporarily remove the probe connected to Channel 2 of the oscilloscope, and adjust the resulting straight line (ground level) at some convenient position on the screen. Reconnect the probe to the output of the integrator, and measure the negative peak voltage (with respect to ground) of the triangle wave, recording your result in Table 34–2.

11. Now measure the time duration for which the triangle-wave signal is negative (t_1). The peak voltage of a triangle wave that results from integrating a square waveform having a peak voltage V_m is given by

$$V_{out}(\text{peak}) - \frac{V_m t_1}{R_1 C}$$

Compute the expected value of the negative peak voltage, and compare it with the measured value above. Record your results in Table 34–2.

12. Change the time base to 50 μs/division and Channel 2 to 1 V/division. Then adjust the input frequency to 4 kHz. Repeat Steps 10 and 11. You should find that the peak output voltage increases.

13. Now change the input frequency to 100 Hz. Adjust the time base to 2 ms/division and Channel 2 to 5 V/division. What does the output signal look like?

    Notice that the output signal looks like a square wave with a phase shift of 180°. Why?

Below approximately 724 Hz, the circuit ceases to act as an integrator since the reactance of the 0.0022-μF capacitor is now greater than that of the 100-kΩ resistor (R_S). Below this frequency, the circuit functions like that of an inverting amplifier having a voltage gain of $-R_S/R_1$.

14. Measure the peak-to-peak output voltage and determine the voltage gain, recording your values in Table 34–2. How does the voltage gain compare to that of an inverting amplifier?

WHAT YOU HAVE DONE

This experiment demonstrated the operation of the differentiator and integrator using 741 operational amplifiers. The differentiator is a circuit that calculates the instantaneous slope of the line at every point on a waveform. On the other hand, the integrator generates a signal that is proportional to the accumulative "area underneath the curve" of a given waveform. Besides these mathematical operations, differentiators and integrators are often used as signal processing circuits because of their ability to change the shapes of their input signals.

Name ______________________ Date ____________

OP-AMP DIFFERENTIATOR AND INTEGRATOR

OBJECTIVES/PURPOSE:

SCHEMATIC DIAGRAM:

Name _____   Date _____

DATA FOR EXPERIMENT 34

TABLE 34–1 *Op-amp differentiator.*

Input Frequency	Measured Peak Output	Expected Peak Output	% Error
400 Hz			
1 kHz			
30 kHz			

TABLE 34–2 *Op-amp integrator.*

Input Frequency	Measured Peak Output	Expected Peak Output	% Error
10 kHz			
4 kHz			
100 Hz			

RESULTS AND CONCLUSIONS:

REVIEW QUESTIONS FOR EXPERIMENT 34

1. The maximum frequency below which the circuit of Figure 34–1A acts as a differentiator is approximately
   (a) 3 kHz **(b)** 3.3 kHz **(c)** 3.6 kHz **(d)** 15 kHz   (   )
2. When the circuit of Figure 34–1A is acting as an amplifier, the voltage gain is
   (a) −10 **(b)** −1 **(c)** 1 **(d)** 10   (   )
3. The minimum frequency above which the circuit of Figure 34–1B acts as an integrator is approximately
   (a) 720 Hz **(b)** 3 kHz **(c)** 1.7 kHz **(d)** 3.4 kHz   (   )
4. A 2-kHz triangle waveform is applied to the circuit of Figure 34–1A. The output signal then looks like a
   (a) triangle waveform with 0° phase shift
   (b) triangle waveform with 180° phase shift
   (c) square wave with 0° phase shift
   (d) square wave with 180° phase shift   (   )
5. A 2-kHz square wave is applied to the circuit of Figure 34–1B. The output signal then looks like a
   (a) triangle waveform with 0° phase shift
   (b) triangle waveform with 180° phase shift
   (c) square wave with 0° phase shift
   (d) square wave with 180° phase shift   (   )

NOTES

35

THE BUTTERWORTH 2ND-ORDER LOW-PASS ACTIVE FILTER

PURPOSE AND BACKGROUND

The purpose of this experiment is to demonstrate the operation and characteristics of a Butterworth Sallen and Key 2nd-order low-pass active filter. A Butterworth low-pass filter passes all signals with frequencies below its cutoff frequency with a constant, or *maximally flat,* passband gain. The cutoff frequency is also referred to as the *critical corner, break,* or *3-dB frequency.* Above this frequency, the input signal is attenuated at a rate of -12 dB/octave, a rate that is equivalent to -40 dB/decade for such a 2nd-order filter. Because of the component values used in this experiment, the passband voltage gain is ideally fixed at 1.586 (4 dB), although other arrangements allowing other higher passband gains are possible. In addition, for input frequencies well below the cutoff frequency, there is no phase shift between input and output signals.

Text Reference: 16–3, Active Low-Pass Filters.

REQUIRED PARTS AND EQUIPMENT

Resistors (1/4 W):
- ☐ Two 6.8 kΩ
- ☐ 27 kΩ
- ☐ 47 kΩ
- ☐ Two 0.033-μF capacitors

- ☐ 741 op-amp (8-pin mini-DIP)
- ☐ Two 0–15 V dc power supplies
- ☐ Signal generator
- ☐ Dual trace oscilloscope
- ☐ Breadboarding socket

317

USEFUL FORMULAS

(1) $R_1 = R_2$ and $C_1 = C_2$

(2) $R_B = 0.586\, R_A$

Cutoff frequency

(3) $f_c = \dfrac{1}{2\pi R_1 C_1}$

dB frequency response

(4) $A_{dB} = 20 \log \left(\dfrac{V_{out}}{V_{in}} \right)$

(5) $A_{dB} = 20 \log \left(\dfrac{1.586}{\left[1 + \left(\dfrac{f_{in}}{f_c} \right)^4 \right]^{1/2}} \right)$

PROCEDURE

1. Wire the circuit shown in the schematic diagram of Figure 35–1.

2. Set your oscilloscope for the following approximate settings:

    Channels 1 and 2: 0.2 V/division, ac coupling
    Time base: 1 ms/division

3. Apply power to the breadboard, and adjust the input signal voltage to 1V peak-to-peak at a frequency of 100 Hz. You should make this voltage setting as accurate as possible.

4. With the resistor and capacitor values used in this circuit, what do you expect the cutoff frequency to be?

    From the formula for cutoff frequency, the cutoff frequency is approximately 710 Hz.

5. With the input frequency set at 100 Hz, what is the peak-to-peak output voltage?

    You should find that the output voltage is larger than the input. You should also observe that both the input and output signals are essentially in-phase.

6. Now vary the generator's frequency (f_{in}), keeping the input voltage constant at 1 V peak-to-peak in order to complete the required data in Table 35–1. At the higher frequencies, you may have to increase the input voltage to obtain a measurable output level. Using the dB frequency response formula (Equation 5), calculate the expected dB response using a cutoff frequency of 710 Hz. Then plot both your experimental and your expected results on the blank graph provided for this purpose.

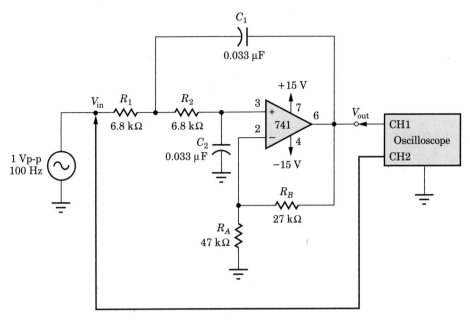

FIGURE 35–1 *Schematic diagram of circuit.*

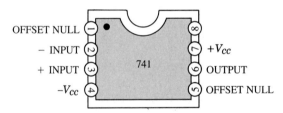

FIGURE 35–2 *Pin diagram of 741 op-amp.*

7. From your plotted results, you should find that both are parallel. How closely they are parallel will depend on how close the actual resistor and capacitor values are to the values shown in the schematic diagram.

8. Notice that the filter's gain at low frequencies is essentially constant up to some point (that is, the passband), after which it decreases at a linear rate with increasing input frequency. This linear decrease in gain as a function of frequency is termed the *roll-off*. To determine the roll-off of your filter, you must determine the slope of a line. Using the data in Table 35–1, subtract the decibel gain measured at 1 kHz from that measured at 10 kHz. The frequency difference from 1 kHz to 10 kHz is *one decade* (that is, a factor of 10). Consequently, the roll-off is the difference in the decibel gain over a one-decade frequency range. From your measurements, what is the filter's roll-off, and how does it compare with what you should expect for a Butterworth 2nd-order low-pass filter?

You should find that the low-pass filter's roll-off is nearly -40 dB/decade, or -12 dB/octave.

9. The filter's cutoff frequency is the frequency at which the dB frequency response is 3 dB *less* than the dB passband gain. This value is equivalent to an output voltage that is 0.707 times the input voltage of the filter. From your graph, estimate the filter's cutoff frequency, and compare it with the value calculated in Step 4.

You should have estimated a critical frequency of approximately 710 Hz. In this case, the passband gain is 1.586, or 4 dB, so the critical frequency occurs when the filter's dB response is 4 dB -3 dB, or $+1$ dB.

WHAT YOU HAVE DONE

This experiment demonstrated operation and characteristics of a Butterworth Sallen and Key 2nd-order low-pass active filter. This filter passed all signals with frequencies below its critical frequency with a relatively constant passband gain. Above the critical frequency, the input signal is attenuated at a linear rate. In this experiment, the following parameters were measured: passband gain, critical frequency, and roll-off. The filter's frequency response was graphed from the measured data.

Name ______________________________ Date ____________

THE BUTTERWORTH 2ND-ORDER LOW-PASS ACTIVE FILTER

OBJECTIVES/PURPOSE:

SCHEMATIC DIAGRAM:

DATA FOR EXPERIMENT 35

TABLE 35–1

Input Frequency (Hz)	V_{in}	V_{out}	V_{out}/V_{in}	Experimental dB Gain	Expected dB Gain*
100	1 V				
200					
300					
400					
500					
600					
800					
1000					
2000					
4000					
5000					
6000					
8000					
10,000					

*Using a cutoff frequency of 710 kHz, for simplicity.

Name _____ Date _____

DATA FOR EXPERIMENT 35

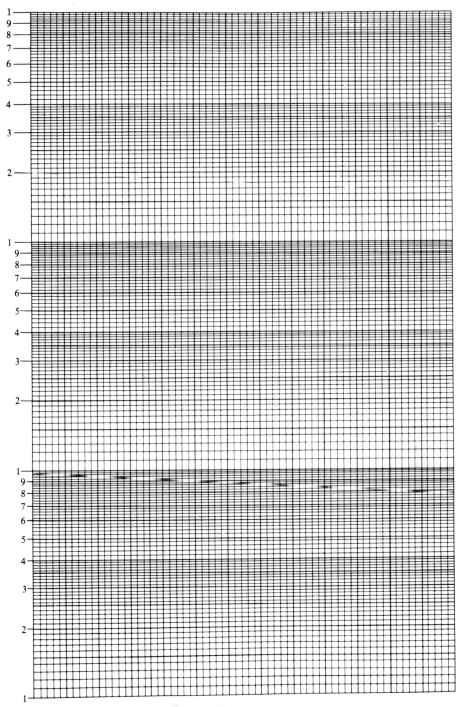

Name _____ Date _____

RESULTS AND CONCLUSIONS:

REVIEW QUESTIONS FOR EXPERIMENT 35

1. For the low-pass filter circuit of Figure 35–1, if R_1 and R_2 are 10 kΩ, the critical frequency is approximately
   (a) 240 Hz   **(b)** 340 Hz   **(c)** 480 Hz   **(d)** 600 Hz   (   )
2. Within the filter's passband, the voltage gain is approximately
   (a) 1   **(b)** 1.5   **(c)** 5   **(d)** 10   (   )
3. Within the filter's passband, the output signal is out-of-phase with the input signal by approximately
   (a) 0°   **(b)** 45°   **(c)** 90°   **(d)** 180°   (   )
4. Beyond the filter's cutoff frequency, the response varies linearly at a rate of
   (a) −6 dB/octave   **(b)** −12 dB/octave
   (c) −18 dB/octave   **(d)** −20 dB/decade   (   )
5. If the cutoff frequency is 500 Hz and the input signal to the filter has a frequency of 3000 Hz, the dB response is
   (a) −3 dB   **(b)** −16 dB   **(c)** −27 dB   **(d)** −62 dB   (   )

324

36

THE BUTTERWORTH 2ND-ORDER HIGH-PASS ACTIVE FILTER

PURPOSE AND BACKGROUND

The purpose of this experiment is to demonstrate the operation and characteristics of a Butterworth 2nd-order Sallen and Key high-pass active filter. A Butterworth high-pass filter has an operation that is opposite that of a low-pass filter. That is, a high-pass filter passes all signals with frequencies *above* its cutoff frequency with a constant, or *maximally flat,* passband gain. Below this frequency, the input signal is attenuated at a rate of 12 dB/octave, or 40 dB/decade for such a 2nd-order filter. Because of the component values used in this experiment, the passband voltage gain is fixed at 1.586 (4 dB), although arrangements allowing other higher passband gains are possible. In addition, for frequencies well above the cutoff frequency, there is no phase shift between input and output signals.

Text Reference: 16–4, Active High-Pass Filters.

REQUIRED PARTS AND EQUIPMENT

Resistors (1/4 W):
- ☐ Two 6.8 kΩ
- ☐ 27 kΩ
- ☐ 47 kΩ
- ☐ Two 0.0047-μF capacitors

- ☐ 741 op-amp (8-pin mini-DIP)
- ☐ Two 0–15 V dc power supplies
- ☐ Signal generator
- ☐ Dual trace oscilloscope
- ☐ Breadboarding socket

325

USEFUL FORMULAS

(1) $R_1 = R_2$ and $C_1 = C_2$

(2) $R_B = 0.586\, R_A$

Cutoff frequency

(3) $f_c = \dfrac{1}{2\pi R_1 C_1}$

dB frequency response

(4) $A_{dB} = 20 \log \left(\dfrac{V_{out}}{V_{in}} \right)$

(5) $A_{dB} = 20 \log \left(\dfrac{1.586}{\left[1 + \left(\dfrac{f_c}{f_{in}} \right)^4 \right]^{1/2}} \right)$

PROCEDURE

1. Wire the circuit shown in the schematic diagram of Figure 36–1.
2. Set your oscilloscope for the following approximate settings:

   Channels 1 and 2: 0.2 V/division, ac coupling
   Time base: 0.2 ms/division

3. Apply power to the breadboard, and adjust the input signal voltage to 1 V peak-to-peak at a frequency of 10 kHz. You should make this voltage setting as accurate as possible.
4. With the resistor and capacitor values used in this circuit, what do you expect the cutoff frequency to be?

   From the formula for the cutoff frequency, the cutoff frequency is approximately 5 kHz.
5. With the input frequency set at 10 kHz, what is the peak-to-peak output voltage?

   You should find that the output voltage is larger than the input. You should also observe that both the input and output signals are essentially in-phase.
6. Now vary the generator's frequency (f_{in}), keeping the input voltage constant at 1 V peak-to-peak in order to complete the required data in Table 36–1. You may have to increase the input voltage at the lower frequencies in order to obtain a measurable output level. Using the dB frequency response formula (Equation 5), calculate the expected dB response using a cutoff frequency of 5 kHz. Then plot both your experimental and your expected results on the blank graph provided for this purpose.

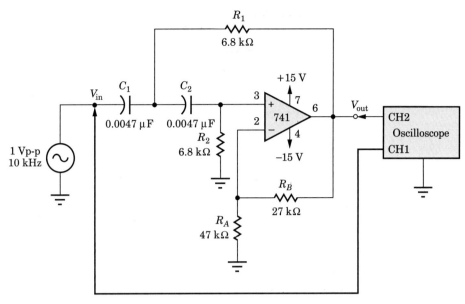

FIGURE 36–1 *Schematic diagram of circuit.*

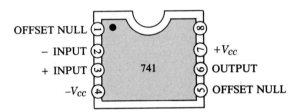

FIGURE 36–2 *Pin diagram of 741 op-amp.*

7. From your plotted results, you should find that both are parallel. How closely they are parallel will depend on how close the actual resistor and capacitor values are to the values shown in the schematic diagram.

8. Notice that the filter's gain at high frequencies is essentially constant from some point (that is, the *passband*), before which it increases at a linear rate with increasing input frequency. This linear increase in gain as a function of frequency is termed the *roll-off*. To determine the roll-off of your filter, you must determine the slope of a line. Using the data in Table 36–1, subtract the dB gain measured at 100Hz from that measured at 1 kHz. The frequency difference from 100 Hz to 10 kHz is *one decade* (that is, a factor of 10). Consequently, the roll-off is the difference in the dB gain over a one-decade frequency range. From your measurements, what is the filter's roll-off, and how does it compare with what you should expect for a Butterworth 2nd-order high-pass filter?

327

You should find that the high-pass filter's roll-off is nearly 40 dB/decade, or 12 dB/octave.

9. The filter's cutoff frequency is the frequency at which the dB frequency response is 3 dB *less* than the dB passband gain. This value is equivalent to an output voltage that is 0.707 times the input voltage of the filter. From your graph, estimate the filter's cutoff frequency, and compare it with the value calculated in Step 4.

   You should have estimated a critical frequency of approximately 5 kHz. In this case, the ideal passband gain is 1.586, or 4 dB, so the critical frequency occurs when the filter's dB response is 4 dB − 3 dB, or +1 dB.

WHAT YOU HAVE DONE

This experiment demonstrated operation and characteristics of a Butterworth Sallen and Key 2nd-order high-pass active filter. This filter passed all signals with frequencies above its critical frequency with a relatively constant passband gain. Below the critical frequency, the input signal is attenuated at a linear rate. In this experiment, the following parameters were measured: passband gain, critical frequency, and roll-off. The filter's frequency response was graphed from the measured data.

Name ______________________________ Date ____________

THE BUTTERWORTH 2ND-ORDER HIGH-PASS ACTIVE FILTER

OBJECTIVES/PURPOSE:

SCHEMATIC DIAGRAM:

DATA FOR EXPERIMENT 36

TABLE 36–1

Input Frequency (Hz)	V_{in}	V_{out}	V_{out}/V_{in}	Experimental dB Gain	Expected dB Gain*
100	1 V				
200					
300					
400					
500					
600					
800					
1000					
2000					
4000					
5000					
6000					
8000					
10,000					

*Using a cutoff frequency of 5 kHz, for simplicity.

330

DATA FOR EXPERIMENT 36

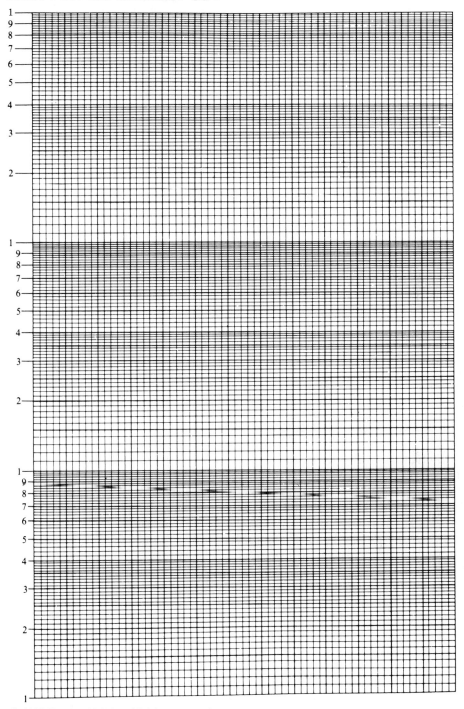

RESULTS AND CONCLUSIONS:

REVIEW QUESTIONS FOR EXPERIMENT 36

1. For the high-pass filter circuit of Figure 36–1, if C_1 and C_2 are 0.001 μF, the critical frequency is approximately
   (a) 8 kHz   **(b)** 12 kHz   **(c)** 16 kHz   **(d)** 23 kHz   (   )
2. Within the filter's passband, the voltage gain is approximately
   (a) 1   **(b)** 1.5   **(c)** 5   **(d)** 10   (   )
3. Within the filter's passband, the output signal is out-of-phase with the input signal by approximately
   (a) 0°   **(b)** 45°   **(c)** 90°   **(d)** 180°   (   )
4. Below the filter's cutoff frequency, the response varies linearly at a rate of
   (a) 6 dB/octave   **(b)** 20 dB/decade
   (c) 40 dB/decade   **(d)** 60 dB/decade   (   )
5. If the cutoff frequency is 2000 Hz and the input signal to the filter has a frequency of 100 Hz, the dB response is
   (a) −3 dB   **(b)** −22 dB   **(c)** −52 dB   **(d)** −104 dB   (   )

332

37

THE ACTIVE BAND-PASS
FILTER

PURPOSE AND BACKGROUND

The purpose of this experiment is to demonstrate the operation and characteristics of a multiple-feedback active band-pass filter. Band-pass filters pass all input signal frequencies within a given range, called the *bandwidth,* while rejecting those frequencies outside this range. The bandwidth encloses a single frequency at which the output voltage is a maximum, called the *center frequency.*

The multiple-feedback band-pass filter is only one of a number of possible band-pass filter circuits which, unlike the "twin-T" band-pass filter, enable one to specify individually the center frequency (f_{10}), gain (A_v), and quality factor (Q). Because of its simplicity, it is limited for Qs less than 10.

Text Reference: 16–5, Active Band-Pass Filters.

REQUIRED PARTS AND EQUIPMENT

Resistors (1/4 W):
- [ ] 1.5 kΩ
- [ ] 2.7 kΩ
- [ ] 68 kΩ
- [ ] 180 kΩ
- [ ] Two 0.01-μF capacitors

- [ ] 741 op-amp (8-pin mini-DIP)
- [ ] Two 0–15 V dc power supplies
- [ ] Signal generator
- [ ] Dual trace oscilloscope
- [ ] Breadboarding socket

333

USEFUL FORMULAS

Center frequency

$$(1) \quad f_0 = \frac{1}{2\pi C}\left(\frac{R_1 + R_1}{R_1 R_2 R_3}\right)^{1/2}$$

where $R_1 = \dfrac{Q}{2\pi f_0 A_0 C}$

$R_2 = \dfrac{Q}{2\pi f_0 C (2Q^2 - G_0)}$

$R_3 = \dfrac{Q}{\pi f_0 C}$

Center frequency voltage gain

$$(2) \quad A_0 = \frac{R_3}{2R_1}$$

where A_0 must be less than $2Q^2$

Shifting the center frequency with constant center frequency gain and bandwidth

$$(3) \quad R_2' = R_2 \left(\frac{f_0}{f_0'}\right)^2$$

Center frequency from upper and lower 3-dB frequencies

$$(4) \quad f_0 = (f_L f_H)^{1/2}$$

Quality factor

$$(5) \quad Q = \frac{f_0}{f_H - f_L}$$

PROCEDURE

1. Wire the circuit shown in the schematic diagram of Figure 37–1, and set your oscilloscope for the following approximate settings:

   Channels 1 and 2: 0.2 V/division, ac coupling
   Time base: 0.2 ms/division

2. Apply power to the breadboard, and adjust the output of the signal generator at 1 V peak-to-peak at a frequency of 1 kHz.

3. Now vary the signal generator's frequency to the point at which the output voltage of the filter, as displayed on Channel 2 of the oscilloscope, reaches its *maximum* peak-to-peak amplitude. Measure this peak-to-peak output voltage, and then determine the center frequency voltage gain V_{out}/V_{in}, recording your data

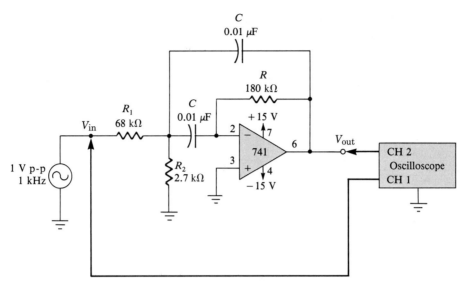

FIGURE 37–1 *Schematic diagram of circuit.*

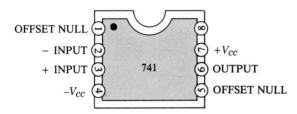

FIGURE 37–2 *Pin diagram of 741 op-amp.*

in Table 37–1. How does the measured voltage gain compare with the expected value (Equation 2)?

The measured center frequency voltage gain, which is based on resistors R_1 and R_3, should be about 1.32. If your value is 10% or more off from this value, either you are not at the filter's center frequency, as evidenced by a maximum output voltage, or the resistors you are using are significantly different from their rated values. You should also observe that the input and output waveforms are *exactly* 180° out-of-phase at this center frequency. Such is the case because the input signal eventually is connected to the op-amp's *inverting* input so that the output signal will be inverted from, or 180° out-of-phase with, the input signal.

4. Using your oscilloscope, determine the filter's output frequency without disturbing the frequency setting of the signal generator, recording your result in Table 37–1. How does this value compare with the expected value (Equation 1)?

The band-pass filter's center frequency is based on the values of both capacitors and all three resistors, and it should be near 737 Hz.

5. Now determine the filter's bandwidth by measuring both the upper and lower 3-dB frequencies at which the peak-to-peak output voltage drops to 0.707 times the value at the center frequency. To do this easily, you should set the signal generator first to the filter's center frequency. Then, without changing the output frequency, adjust the signal generator's output voltage so that the output voltage of the filter is 1.0 V. Make this setting as accurate as possible.

   Then decrease the signal generator's frequency, and stop at the point at which the output voltage drops to 0.71 V peak-to-peak (1.0 V × 0.707 ≃ 0.71). Determine the frequency at this point, called the *lower 3-dB frequency* (f_L), and record your result in Table 37–2.

6. Continue to decrease the input frequency. Does the output voltage increase or decrease?

   Notice that the peak-to-peak output voltage of the band-pass filter *decreases* as the input frequency moves away from the filter's center frequency.

7. Now increase the signal generator's frequency beyond the center frequency, and stop at the point at which the filter's peak-to-peak output voltage is again 0.71 V. Determine the frequency at this point, called the *upper 3-dB frequency* (f_H), and record this result in Table 37–2.

8. Subtract the lower 3-dB frequency from the upper 3-dB frequency, obtaining the *3-dB bandwidth* of the filter. Record your result in Table 37–2. Using this bandwidth and the center frequency experimentally found in Step 4, calculate the filter's Q, or *quality factor,* and record your result in Table 37–2.

   Within 10%, you should determine a filter Q of 4.17. If not, repeat Steps 3 through 7, carefully measuring the voltages and frequencies.

9. From the two measured 3-dB frequencies, you can determine the filter's center frequency by taking the *geometric average:*

$$f_0 = (f_L f_H)^{1/2}$$

How does your result obtained from this equation compare with the value you determined in Step 4?

10. Disconnect both the power and the signal leads from the breadboard, and replace the 2.7-kΩ resistor (R_2) with a 1.5-kΩ resistor. Connect the power and signal generator to the breadboard.

11. Repeat steps 3 through 9 to determine the filter's center frequency voltage gain (A_0), center frequency (f_0), bandwidth, and Q. Record your results in Table 37–3.

12. When the value of Resistor R_2 is changed, how does the new center frequency that you determined in Table 37–3 compare with the expected value obtained from Equation 3 in the "Useful Formulas" section of this experiment?

    If you have performed this experiment correctly, you should find that when resistor R_2 is changed, the bandwidth and the center frequency gain remain the same, while the filter's center frequency is *inversely* proportional to the value of R_2. For example, if R_2 changes from 2.7 kΩ to 1.5 kΩ, the new center frequency should be

$$f_0' = (737 \text{ Hz}) \left(\frac{2.7 \text{ k}\Omega}{1.5 \text{ k}\Omega} \right)^{1/2} = 988 \text{ Hz}$$

Since the center frequency changes, Q also changes.

13. Set the input voltage to the filter at 1 V peak-to-peak, and vary the signal generator's frequency according to Table 37–4. Then plot the dB gain response for all measured frequencies on the blank graph provided for this purpose. From this graph, you should be able to estimate the filter's center frequency, bandwidth, and Q, and these values should compare favorably with those in Table 37–3.

WHAT YOU HAVE DONE

This experiment demonstrated operation and characteristics of a multiple-feedback band-pass active filter. This filter passed all signals within a given range about the filter's center frequency while rejecting those frequencies outside this range. In this experiment, the following parameters were measured: center frequency gain, center frequency, bandwidth, and Q. In addition, it was shown how to vary the filter's center frequency with a single resistor while keeping the bandwidth and center frequency gain constant. The filter's frequency response was graphed from the measured data.

NOTES

Name ______________________________ Date ____________

THE ACTIVE BAND-PASS FILTER

OBJECTIVES/PURPOSE:

SCHEMATIC DIAGRAM:

DATA FOR EXPERIMENT 37

TABLE 37–1

Input voltage, V_{in}	V
Output voltage, V_{out}	V
Center frequency voltage gain, A_0	
Center frequency, f_0	Hz

TABLE 37–2

Lower 3-dB frequency, f_L	Hz
Upper 3-dB frequency, f_H	Hz
3-dB bandwidth, BW	Hz
Quality factor, Q	

TABLE 37–3

Input voltage, V_{in}	V
Output voltage, V_{out}	V
Center frequency voltage gain, A_0	
Center frequency, f_0	Hz
Lower 3-dB frequency	Hz
Upper 3-dB frequency	Hz
3-dB bandwidth, BW	Hz
Quality factor, Q	

TABLE 37–4

Input Frequency (Hz)	V_{in}	V_{out}	V_{out}/V_{in}	Measured dB Gain
100	1 V			
200				
400				
600				
800				
1000				
2000				
4000				
6000				
8000				
10,000				

NOTES

DATA FOR EXPERIMENT 37

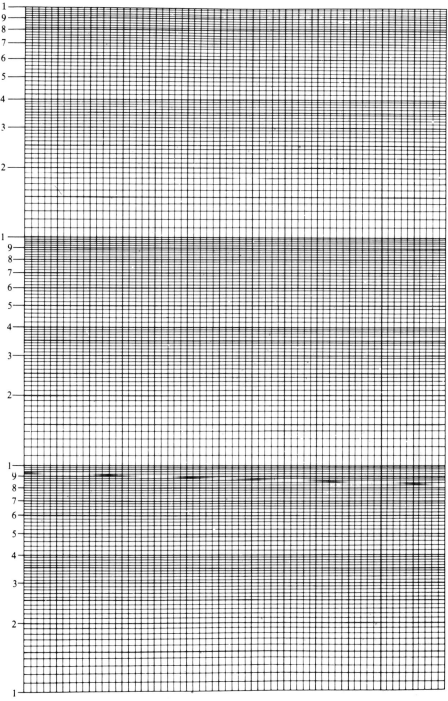

Name _____   Date _____

RESULTS AND CONCLUSIONS:

REVIEW QUESTIONS FOR EXPERIMENT 37

1. The center frequency for the band-pass filter circuit of Figure 37–1 with the components shown is approximately
   (a) 350 Hz   (b) 500 Hz   (c) 625 Hz   (d) 740 Hz   (   )

2. The Q for the circuit of Figure 37–1 with the components shown is approximately
   (a) 2   (b) 4   (c) 6   (d) 8   (   )

3. At the filter's center frequency, the output signal is out-of-phase with the input by
   (a) 0°   (b) 45°   (c) 90°   (d) 180°   (   )

4. If the center frequency is 1000 Hz and the Q is 5, the 3-dB bandwidth is
   (a) 100 Hz   (b) 200 Hz   (c) 300 Hz   (d) 500 Hz   (   )

5. For the circuit of Figure 37–1, if R_2 is changed to a lower value,
   (a) the center frequency gain increases
   (b) the bandwidth decreases
   (c) the center frequency increases
   (d) all of the above   (   )

344

38

THE ACTIVE
BAND-STOP FILTER

PURPOSE AND BACKGROUND

The purpose of this experiment is to demonstrate the operation and characteristics of an active notch, or band-stop, filter. The operation of a band-stop filter is opposite that of a band-pass filter in that the notch filter rejects all input signal frequencies within a given range, called the *bandwidth,* while passing those frequencies outside this range. The bandwidth encloses a single frequency where the output voltage is a minimum, called the *center, notch,* or *null frequency.*

This experiment uses a multiple-feedback band-pass filter with a two-input summing amplifier to create a band-stop filter. This filter is only one of a number of possible band-stop filter circuits.

Text Reference: 16–6, Active Band-Stop Filters.

REQUIRED PARTS AND EQUIPMENT

Resistors (1/4 W):
- ☐ 2.7 kΩ
- ☐ Two 12 kΩ
- ☐ 15 kΩ
- ☐ 68 kΩ
- ☐ 180 kΩ
- ☐ Two 0.01-μF capacitors

- ☐ Two 741 op-amps (8-pin mini-DIP)
- ☐ Two 0–15 V dc power supplies
- ☐ Signal generator
- ☐ Dual trace oscilloscope
- ☐ Breadboarding socket

USEFUL FORMULAS

Notch frequency

$$(1) \quad f_0 = \frac{1}{2\pi C} \left(\frac{R_1 + R_2}{R_1 R_2 R_3} \right)^{1/2}$$

where $R_1 = \dfrac{Q}{2\pi f_0 A_0 C}$

$R_2 = \dfrac{Q}{2\pi f_0 C(2Q^2 - A_0)}$

$R_3 = \dfrac{Q}{\pi f_0 C}$

For unity passband voltage gain

$$(2) \quad A_0 = \frac{R_3}{2R_1}$$

where $R_4 = A_0 R_5$ and $R_6 = R_5$

Quality factor

$$(3) \quad Q = \frac{f_0}{f_H - f_L}$$

PROCEDURE

1. Wire the circuit shown in the schematic diagram of Figure 38–1, and set your oscilloscope for the following approximate settings:

   Channels 1 and 2: 0.2 V/division, ac coupling
   Time base: 0.2 ms/division

2. Apply power to the breadboard, and adjust the output of the signal generator at 1 V peak-to-peak at a frequency of 1 kHz.

3. Now slowly decrease the signal generator's frequency to the point at which the output voltage of the filter, as displayed on Channel 2 of the oscilloscope, reaches its *minimum* peak-to-peak amplitude. You will need to increase the sensitivity of Channel 2 to do this. Measure the peak-to-peak output voltage. Then determine the center frequency voltage gain V_{out}/V_{in} (in decibels), sometimes called the *notch depth* or *depth of null*, and record your data in Table 38–1. You should measure a notch depth of at least –25 dB.

   If you do not have a notch depth of at least –25 dB, you are probably not at the filter's center frequency, as evidenced by a minimum output voltage. Also, observe that the input and output waveform are *exactly* in-phase at this center frequency.

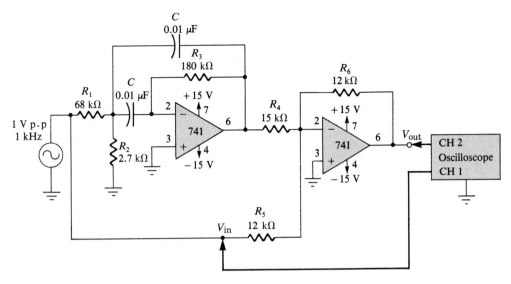

FIGURE 38–1 *Schematic diagram of circuit.*

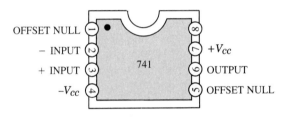

FIGURE 38–2 *Pin diagram of 741 op-amp.*

4. Using your oscilloscope, determine the filter's center frequency without disturbing the frequency setting of the signal generator, recording your result in Table 38–1. How does this value compare with the expected value (Equation 1)?

  The notch filter's center frequency is based on the values of both capacitors and all three resistors, and it should be near 737 Hz.

5. Continue to decrease the input frequency. Does the output voltage increase or decrease?

  Notice that the peak-to-peak output voltage of the notch filter *increases* as the input frequency moves away from the filter's center frequency. Eventually, the output voltage remains essentially constant (the passband).

6. Set the frequency of the signal generator at 100 Hz and the oscilloscope's Channel 2 sensitivity to 0.2 V/division. Measure the peak-to-peak output voltage of the filter, and determine the passband voltage gain, recording your results in Table 38–2.

7. Now determine the filter's bandwidth by measuring both the upper and the lower 3-dB frequencies at which the peak-to-peak output voltage drops to 0.707 times the value of the output voltage in the filter's passband. To do this easily, first you should set the signal generator at 100 Hz. Then, without changing the output frequency, adjust the signal generator's output voltage so that the output voltage of the filter is 1.0 V. Make this setting as accurate as possible.

   Then decrease the signal generator's frequency, and stop at the point at which the output voltage drops to 0.71 V peak-to-peak (1.0 V × 0.707 ≃ 0.71). Determine the frequency at this point, called the *lower 3-dB frequency* (f_L), and record your result in Table 38–2.

8. Continue to decrease the input frequency. Does the output voltage increase or decrease?

   Notice that the peak-to-peak output voltage of the notch filter *decreases* as the input frequency moves toward the filter's center frequency.

9. Now increase the signal generator's frequency beyond the center frequency, and stop at the point at which the filter's peak-to-peak output voltage is again 0.71 V. Determine the frequency at this point, called the *upper 3-dB frequency* (f_H), and record this result in Table 38–2.

10. Subtract the lower 3-dB frequency from the upper 3-dB frequency, obtaining the *3-db bandwidth* of the filter. Record your result in Table 38–2. Using this bandwidth and the center frequency found experimentally in Step 4, calculate the filter's *Q*, or *quality factor*, and record your result in Table 38–2.

    Within 10 percent, you should determine a filter *Q* of 4.17. If not, repeat Steps 3 through 9, carefully measuring the voltages and frequencies.

11. From the two measured 3-dB frequencies, you can determine the filter's center frequency by taking the *geometric average*:

    $$f_0 = (f_L f_H)^{1/2}$$

    How does your result obtained from this equation compare with the value you determined in Step 4?

12. Vary the signal generator's frequency according to Table 38–3, and plot the dB gain for all measured frequencies on the blank graph provided for this purpose. From this graph, you should be able to estimate the filter's center frequency, bandwidth, and *Q*.

WHAT YOU HAVE DONE

This experiment demonstrated operation and characteristics of a multiple-feedback band-stop (notch) active filter. This filter rejected all signals within a given range about the filter's center frequency while passing those frequencies outside this range. In this experiment, the following parameters were measured: passband gain, center (notch) frequency, bandwidth, and Q. The filter's frequency response was graphed from the measured data.

NOTES

Name _____   Date _____

THE ACTIVE BAND-STOP FILTER

OBJECTIVES/PURPOSE:

SCHEMATIC DIAGRAM:

DATA FOR EXPERIMENT 38

TABLE 38–1

Input voltage, V_{in}	V
Output voltage, V_{out}	V
Notch Depth	dB
Center frequency, f_0	Hz

TABLE 38–2

Pass-band voltage gain, A_0	
Lower 3-dB frequency, f_L	Hz
Upper 3-dB frequency, f_H	Hz
3-dB bandwidth, BW	Hz
Quality factor, Q	

Name _____   Date _____

TABLE 38–3

Input Frequency (Hz)	V_{in}	V_{out}	V_{out}/V_{in}	Measured dB Gain
100	1 V			
200				
400				
600				
700				
750				
800				
1000				
2000				
4000				
6000				

NOTES

DATA FOR EXPERIMENT 38

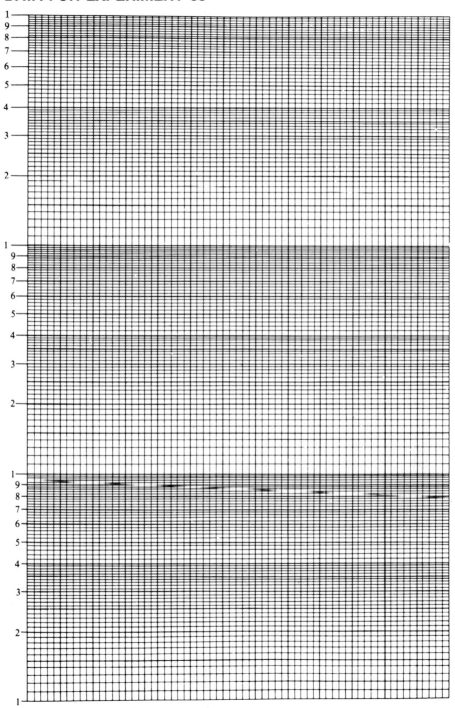

Name _____ Date _____

RESULTS AND CONCLUSIONS:

REVIEW QUESTIONS FOR EXPERIMENT 38

1. The notch frequency of the filter circuit of Figure 38–1 with the components shown is approximately
   (a) 350 Hz   **(b)** 500 Hz   **(c)** 625 Hz   **(d)** 740 Hz   (   )

2. The Q for the circuit of Figure 38–1 with components shown is approximately
   (a) 2   **(b)** 4   **(c)** 6   **(d)** 8   (   )

3. At the filter's notch frequency, the output signal is out-of-phase with the input by
   (a) 0°   **(b)** 45°   **(c)** 90°   **(d)** 180°   (   )

4. In the filter's passband, the voltage gain is approximately
   (a) 0   **(b)** 1   **(c)** 2   **(d)** 4   (   )

5. From your filter response curve, the maximum rejection of the input signal occurs at
   (a) frequencies within the passband
   (b) the lower 3-dB frequency
   (c) the notch frequency
   (d) the upper 3-dB frequency   (   )

356

39

THE PHASE-SHIFT
OSCILLATOR

PURPOSE AND BACKGROUND

The purpose of this experiment is to demonstrate the design and operation of an op-amp phase-shift oscillator. By providing three RC networks having a total phase shift of 180° as positive feedback to the input of an inverting amplifier, oscillation results. The total phase shift of the amplifier and that of the phase-shift network is 0°, and the loop gain is unity. The sinusoidal output of the oscillator has a peak to-peak voltage equal to the difference between the op-amp positive and negative saturation voltages.

Text Reference: 17–3, Oscillators with *RC* Feedback Circuits.

REQUIRED PARTS AND EQUIPMENT

Resistors (1/4 W):
- ☐ Three 1 kΩ
- ☐ 27 kΩ
- ☐ 5-kΩ potentiometer
- ☐ Three 0.1-μF capacitors

- ☐ 741 op-amp (8-pin mini-DIP)
- ☐ Two 0–15 V dc power supplies
- ☐ Dual trace oscilloscope
- ☐ Breadboarding socket

USEFUL FORMULAS

Output frequency

$$(1) \ f_o = \frac{1}{2\pi RC \sqrt{6}}$$

For oscillation

$$(2) \ R_f/R = 29$$

PROCEDURE

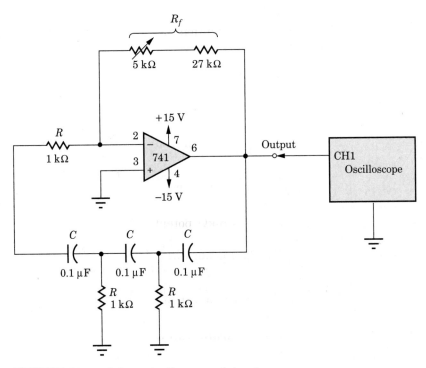

FIGURE 39–1 *Schematic diagram of circuit.*

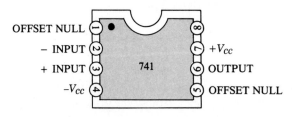

FIGURE 39–2 *Pin diagram of 741 op-amp.*

358

1. Wire the circuit shown in the schematic diagram of Figure 39–1 and set your oscilloscope to the following approximate settings:

   Channel 1: 5 V/division, ac coupling
   Time base: 0.5 ms/division

2. After you have checked all connections, apply the ± 15-V power supply connections to the breadboard.

3. Depending on the setting of the 5-kΩ potentiomenter, the circuit may or may not be oscillating when power is applied. If a sine wave is not displayed on the oscilloscope, carefully adjust the 5-kΩ potentiometer until a sine wave starts to appear on the oscilloscope's display. If you continue to increase the resistance of the potentiometer, you should observe that the peaks of the sine wave become clipped and that the output frequency becomes lower. Adjust the potentiometer to the point at which the circuit just sustains oscillation.

   On the other hand, if a sine wave is seen when power is applied on the breadboard, carefully decrease the resistance of the potentiometer to obtain the best-looking sine wave.

4. Using your oscilloscope's time base set at approximately 0.2 ms/division, measure the output frequency of the phase-shift oscillator, recording your result in Table 39–1. Compare this value with the expected frequency found using Equation 1 given in the Useful Formulas section of this experiment.

5. Disconnect the power from the breadboard and carefully remove and measure the total resistance (R_f) of the 27-kΩ resistor and of the setting of the series 5-kΩ potentiometer that produced oscillation. Record the value in Table 39–1. At the oscillation frequency set by the three RC networks, 1/29 of the output signal is fed back to the input of the op-amp. For the loop gain to be unity, the voltage gain of the inverting amplifier must then be 29, which implies that the feedback resistor R_f must be equal to $29R$. How does the sum of the 27-kΩ resistor and setting of the 5-kΩ potentiometer compare with the 1-kΩ resistors of the phase-shift network?

WHAT YOU HAVE DONE

This experiment demonstrated the design and operation of a phase-shift oscillator using a 741 operational amplifier. This type of oscillator uses three RC networks having a total phase shift of $180°$ at a specific frequency as positive feedback to the input of an inverting amplifier. When the loop gain is 1 by making $R_f = 29R$, the circuit then oscillates.

Name _____ Date _____

THE PHASE SHIFT OSCILLATOR

OBJECTIVES/PURPOSE:

SCHEMATIC DIAGRAM:

DATA FOR EXPERIMENT 39

TABLE 39–1

Parameter	Measured	Expected	% Error
Output Frequency, f_o			
R_f		29 kΩ	

Name _____ Date _____

RESULTS AND CONCLUSIONS:

REVIEW QUESTIONS FOR EXPERIMENT 39

1. The RC networks in the oscillator of Figure 39–1 are a form of
   (a) positive feedback   **(b)** negative feedback   (   )
2. For oscillation, the voltage gain of the amplifier must be
   (a) 1/29   **(b)** 1   **(c)** 29   **(d)** infinite   (   )
3. The output of the oscillator is a
   (a) square wave   **(b)** sine wave   **(c)** triangle wave
   (d) sawtooth   **(e)** rectified sine wave   (   )
4. If the values of the three capacitors are increased, the output
   frequency will
   (a) increase   **(b)** decrease   **(c)** remain the same   (   )
5. If the value of the feedback resistor R_f in Figure 39–1 is made
   less than 29 kΩ,
   (a) the circuit continues oscillating at the same frequency
   (b) the output frequency increases
   (c) the output frequency decreases
   (d) the sine wave changes into a square wave
   (e) the circuit stops oscillating   (   )

NOTES

40

THE 555 TIMER ASTABLE MULTIVIBRATOR

PURPOSE AND BACKGROUND

The purpose of this experiment is to demonstrate the operation of the 555 timer as an astable multivibrator. The 555 timer is an IC device that allows the formation of an astable multivibrator whose output frequency and percent duty cycle can be controlled by only two resistors and a single capacitor. As a practical matter, the output frequency should be kept less than 200 kHz, while the duty cycle can range from approximately 50% to 99%.

Text Reference: 17–6, the 555 Timer As an Oscillator.

REQUIRED PARTS AND EQUIPMENT

Resistors (1/4 W):
- ☐ 1 kΩ
- ☐ 3.3 kΩ
- ☐ 15 kΩ

Capacitors:
- ☐ 0.01 μF
- ☐ 1 μF

- ☐ 555 timer (8-pin mini-DIP)
- ☐ 0–15 V dc power supply
- ☐ Dual trace oscilloscope
- ☐ Breadboarding socket

USEFUL FORMULAS

Output frequency

$$(1)\ f_o = \frac{1.44}{(R_1 + 2R_2)C} = \frac{1}{T}$$

Percent duty cycle

$$(2)\ \%D = \frac{R_1 + R_2}{R_1 + 2R_2} \times 100\% = \frac{t_1}{T} \times 100\%$$

PROCEDURE

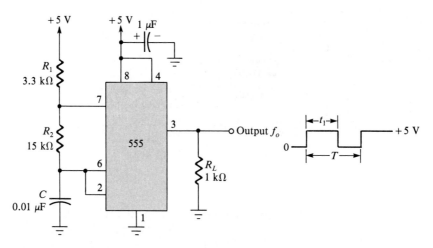

FIGURE 40–1 *Schematic diagram of circuit.*

1. Wire the circuit shown in the schematic diagram of Figure 40–1 using a 5 V supply. Set the oscilloscope to the following approximate settings:

   Channel 1: 2 V/division, dc coupling
   Time base: 0.1 ms/division

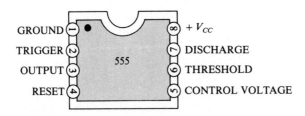

FIGURE 40–2 *Pin diagram of 555 timer.*

2. Apply power to the breadboard. You should see a waveform that switches back and forth between ground and the +5-V supply voltage, similar to that shown in Figure 40–3. Measure the output frequency, and compare it to the value that you would expect based on the values of R_1, R_2, and C (Equation 1). Record your results in Table 40–1.

555 timer output
Pin 3

FIGURE 40–3   *555 timer output, pin 3.*

3. Determine the percent duty cycle of the output waveform of the 555 timer astable multivibrator by taking the ratio of the time that the waveform is at the positive supply voltage to the total time for one cycle. Then multiply this result by 100%. Compare your result with the expected value (Equation 2), and record both in Table 40–1.

4. Disconnect the power from the breadboard, and reverse the 3.3-kΩ and 15-kΩ timing resistors so that the 15-kΩ resistor is now R_1. Again connect power to the breadboard, and compare the measured output frequency with the expected value (Equation 1), recording your results in Table 40–1.

5. As in Step 3, measure the percent duty cycle, comparing it to the expected value (Equation 2), and record your results in Table 40–1.

Observe that if resistor R_2 is much greater than R_1, the percent duty cycle will approach 50%. On the other hand, if R_1 is much larger than R_2, then the percent duty cycle approaches 99%. However, note that when either R_1 or R_2 is changed to adjust the duty cycle, the output frequency of the 555 timer also changes. Consequently, the output frequency and the duty cycle, once set, cannot be adjusted independently.

WHAT YOU HAVE DONE

This experiment demonstrated the operation of the 555 timer as an astable multivibrator and determined what components controlled its output frequency and duty cycle.

Name _____   Date _____

THE 555 TIMER ASTABLE MULTIVIBRATOR

OBJECTIVES/PURPOSE:

SCHEMATIC DIAGRAM:

369

Name _____ Date _____

DATA FOR EXPERIMENT 40

TABLE 40–1

Component Values	Output Frequency			% Duty Cycle		
	Measured	Calculated	% Error	Measured	Calculated	% Error
Steps 2 and 3: $R_1 = 3.3$ kΩ $R_2 = 15$ kΩ $C = 0.01$ μF						
Steps 4 and 5: $R_1 = 15$ kΩ $R_2 = 3.3$ kΩ $C = 0.01$ μF						

Name _____ Date _____

RESULTS AND CONCLUSIONS:

REVIEW QUESTIONS FOR EXPERIMENT 40

1. For the 555 timer astable multivibrator circuit of Figure 40–1, the output frequency is approximately
    (a) 4.3 kHz   **(b)** 5.5 kHz   **(c)** 6.0 kHz   **(d)** 7.9 kHz   (   )
2. For the 555 timer astable multivibrator circuit of Figure 40–1, the percent output duty cycle is approximately
    (a) 22%   **(b)** 45%   **(c)** 55%   **(d)** 78%   (   )
3. If R_1 is made much larger than R_2, the percent duty cycle approaches
    (a) 0%   **(b)** 50%   **(c)** 99%   (   )
4. If R_2 is made much larger than R_1, the percent duty cycle approaches
    (a) 0%   **(b)** 50%   **(c)** 99%   (   )
5. When either or both of the timing resistors are changed,
    (a) only the output frequency changes
    (b) only the percent duty cycle changes
    (c) both the frequency and the percent duty cycle change
    (d) nothing happens   (   )

NOTES

41

THE PHASE DETECTOR

PURPOSE AND BACKGROUND

The purposes of this experiment are (1) to demonstrate the operation and characteristics of a phase detector when both input frequencies are the same (that is, phase lock) and (2) to determine its conversion gain. The output of a phase detector has a dc voltage, V_{out}, that is proportional to the phase difference, $\Delta\phi$, of its two input signals. The ratio of the change of this dc output voltage to the corresponding change in phase is called the *conversion gain*, K_ϕ, of the phase detector.

This experiment uses a D-type flip-flop functioning as a phase detector, which is only one of many possible circuits that may be used.

Text Reference: 17–7, The Phase-Locked Loop.

REQUIRED PARTS AND EQUIPMENT

- ☐ 7404 TTL hex inverter
- ☐ 7442 BCD-to-decimal decoder
- ☐ 7474 dual D-type flip-flop
- ☐ 7490 decade counter
- ☐ 5-V dc power supply
- ☐ TTL level signal generator
- ☐ Dual trace oscilloscope
- ☐ VOM or DMM (preferred)
- ☐ Breadboarding socket

373

USEFUL FORMULA

Phase-detector conversion gain

$$K_\phi = \frac{\Delta V_{out}}{\Delta \phi} \quad \text{(volts/radian)}$$

PROCEDURE

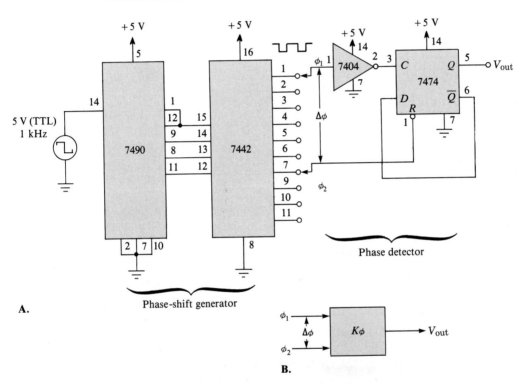

A.

Phase-shift generator

B.

FIGURE 41–1 *Schematic diagram of circuit.*

1. Wire the circuit shown in the schematic diagram of Figure 41–1. The 7490 decade counter and the 7442 decoder make a simple phase-shift generator having ten fixed phase shifts from 0° to 324° in increments of 36°. The 7404 inverter, along with the 7474 flip-flop, make up the phase detector, which has two inputs (pin 1 of the 7404 and pin 1 of the 7474) and a single output (pin 5 of the 7474).

   Set your oscilloscope to the following approximate settings:

   Channels 1 and 2: 2 V/division, ac coupling
   Time base: 0.1 ms/division

2. Initially connect pin 1 of the 7474 flip-flop to pin 1 of the 7442 decoder, and apply power to the breadboard. Since both phase-

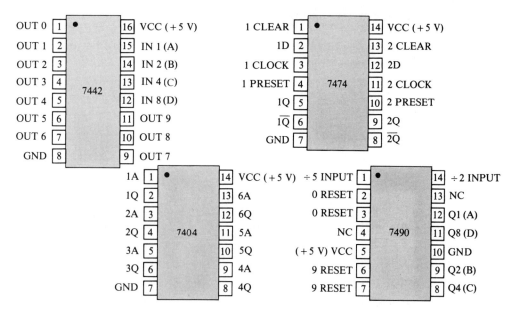

FIGURE 41–2 *Pin diagram of IC devices.*

detector inputs are connected to the same point, the phase shift is naturally 0°. Adjust the output of the square-wave generator so that the signal of Channel 1 occupies exactly one cycle for the ten horizontal divisions (that is, an input frequency of 1 kHz). On the oscilloscope, the two signals should appear as shown in Figure 41–3. Measure the dc output voltage with your VOM or DMM, and record your result in Table 41–1.

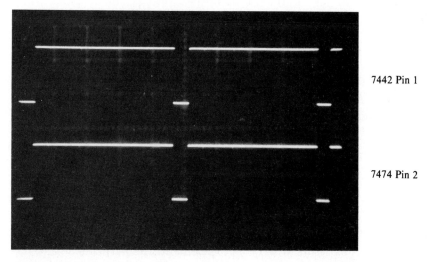

7442 Pin 1

7474 Pin 2

FIGURE 41–3 *0° phase shift.*

3. As directed by Table 41–1, in sequence, connect one input of the phase detector (pin 1 of the 7474 flip-flop) to the different 7442 decoder outputs, while keeping the remaining input of the phase detector connected to pin 1 of the 7442 decoder. At each setting, measure the dc output voltage using a VOM or DMM, and record the results in Table 41–1.

Also, observe the change in phase shift of the two input signals of the phase detector on the oscilloscope each time you increase the amount of phase shift. For example, the two signals having a 72° phase shift are shown in Figure 41–4.

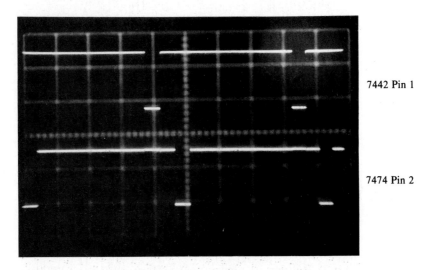

7442 Pin 1

7474 Pin 2

FIGURE 41–4 *72° phase shift.*

4. Now plot your data, the dc output voltage versus phase shift in degrees, on the blank graph provided for this purpose. Your graph should resemble a straight line that increases in voltage as the phase shift increases.

5. From your data and the graph, compute the conversion gain, K_ϕ, in terms of V/radian (that is, the "slope of the line") for this phase detector. For example, if the dc output voltages are 1.0 V and 4.5 V at phase angles of 0° and 324°, respectively, the conversion gain is

$$K_\phi = \frac{\Delta V_{out}}{\Delta \phi}$$

$$= \frac{(4.5 - 1.0V)}{(324° - 0°)}(57.3°/\text{radian})$$

$$K_\phi = 0.619 \text{ V/radian}$$

For comparison, when this experiment was performed, the conversion gain was determined to be 0.594 V/radian.

WHAT YOU HAVE DONE

This experiment demonstrated the operation and characteristics of a simple phase detector. A key component of the phase-locked loop, the output of the phase detector has a dc or average voltage that is proportional to the phase difference of its two input signals. The results of this experiment were graphed and the conversion gain of the phase detector was determined.

NOTES

Name ______________________________ Date ____________

THE PHASE DETECTOR

OBJECTIVES/PURPOSE:

SCHEMATIC DIAGRAM:

DATA FOR EXPERIMENT 41

TABLE 41–1

Phase-Detector Input (7442 Output Pin)	Phase Shift	dc Output Voltage
1	0°	
2	36°	
3	72°	
4	108°	
5	144°	
6	180°	
7	216°	
9	252°	
10	288°	
11	324°	
Phase-detector conversion gain, K_ϕ		V/radian

Name _____ Date _____

DATA FOR EXPERIMENT 41

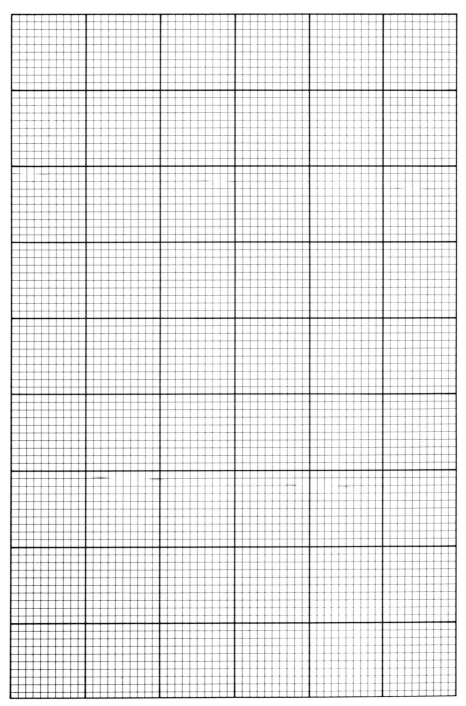

Name _____ Date _____

RESULTS AND CONCLUSIONS:

REVIEW QUESTIONS FOR EXPERIMENT 41

1. The dc output voltage of the D-type flip-flop phase-detector circuit of Figure 41–1
   (a) increases linearly with phase shift
   (b) decreases linearly with phase shift
   (c) depends on the frequencies of the two inputs
   (d) does not change with phase shift   (   )
2. For the circuit in Figure 41–1, the input signals to the phase detector are
   (a) equal in frequency   (b) two different frequencies
   (c) harmonically related (d) none of the above   (   )
3. The phase-detector conversion gain relates
   (a) output and input voltages
   (b) output and input frequencies
   (c) output voltage and input phase shift
   (d) all of the above   (   )
4. If the conversion gain is 2 V/radian and the phase shift is 90°, the change in output voltage of the phase detector is approximately
   (a) 0.3 V   (b) 0.5 V   (c) 1.5 V   (d) 3.1 V   (   )

42

THE 567 PHASE-LOCKED
LOOP TONE DECODER

PURPOSE AND BACKGROUND

The purpose of this experiment is to demonstrate the operation of the 567 phase-locked loop tone decoder. The 567 is an IC phase-locked loop, requiring a minimum of external components to set the VCO frequency up to a maximum of 500 kHz as well as its lock range. It is used primarily to indicate whether a sustained frequency signal or tone is within the loop's capture range. A voltage level equal to the supply voltage indicates that the loop is unlocked, while a zero voltage level indicates a locked system. This experiment uses an LED to indicate whether the phase-locked loop system is locked.

Text Reference: 17–7, The Phase-Locked Loop.

REQUIRED PARTS AND EQUIPMENT

Resistors (1/4 W):
- ☐ 180 Ω
- ☐ 1 kΩ
- ☐ 15 kΩ

Capacitors:
- ☐ Two 0.1 μF
- ☐ 4.7 μF
- ☐ 10 μF

- ☐ 567 phase-locked loop tone decoder
- ☐ LED
- ☐ 5-V power supply
- ☐ Function generator
- ☐ Dual trace oscilloscope
- ☐ Frequency counter
- ☐ Breadboarding socket

383

USEFUL FORMULAS

VCO free-running frequency

(1) $f_0 \simeq \dfrac{1.1}{R_1 C_1}$ (Hz)

(2) $C_2 \geq \dfrac{130}{f_0}$ (μF)

(3) $C_3 > 2C_2$

PROCEDURE

1. Wire the circuit shown in the schematic diagram of Figure 42–1. Set your oscilloscope to the following approximate settings:

   Channel 1: 1 V/division, ac coupling
   Channel 2: 5 V/division, dc coupling
   Time base: 0.5 ms/division

2. Apply power to the breadboard, and adjust the output of the signal generator at approximately 200 Hz with a peak-to-peak voltage of 2 V. The output of the 567 tone decoder should read approximately +5 V, and the LED should be lit, indicating that the phase-locked loop is *unlocked*.

3. Now slowly increase the input frequency until the LED goes out. The output of the 567 should now be approximately zero volts. Measure the input frequency at this point (f_1), and record this value in Table 42–1.

4. Slowly increase the input frequency until the LED is again lit, at which point the output of the 567 will return to approximately +5 V. Measure the input frequency at this point (f_2), and record this value in Table 42–1.

5. Now set the input frequency at 2 kHz. The LEE should be lit, indicating that the loop is unlocked. Slowly decrease the input frequency until the LED becomes unlit. Measure the input frequency at this point (f_3), and record this value in Table 42–1.

6. Slowly decrease the input frequency until the LED again is lit. Measure the input frequency at this point (f_4), and record this value in Table 42–1.

7. Now set the input frequency at approximately 200 Hz. The LED should be lit. Measure the frequency at pin 5 of the 567 tone decoder with your oscilloscope or frequency counter. Since the loop is now unlocked, the frequency at pin 5 is the *free-running VCO frequency (f_0)*. Record this frequency in Table 42–1.

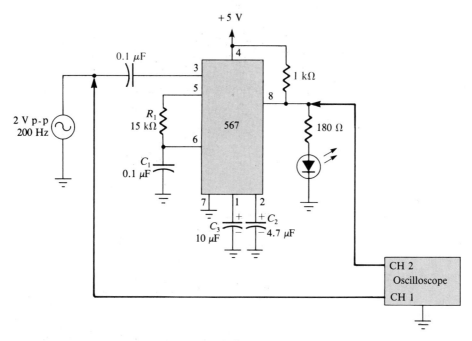

FIGURE 42–1 *Schematic diagram of circuit.*

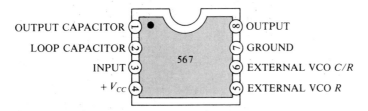

OUTPUT CAPACITOR — 1 8 — OUTPUT
LOOP CAPACITOR — 2 7 — GROUND
INPUT — 3   567   6 — EXTERNAL VCO C/R
$+V_{CC}$ — 4 5 — EXTERNAL VCO R

FIGURE 42–2 *Pin diagram of 567 tone decoder.*

8. From your measurements in Steps 3 through 6, you have determined the range of frequencies at which the 567 tone decoder will capture and then lock. On increasing frequencies,
the phase-locked loop will capture and lock at f_1 and will stay
locked until the input frequency exceeds f_2. On decreasing frequencies, lock will occur at f_3 and will remain until the input
frequency is less than f_4. When the loop is locked, the VCO frequency equals the input frequency. When the loop is unlocked,
the VCO frequency equals its free-running frequency f_0, which
is determined by resistor R_1 and by C_1. From the known values of R_1 and C_1, calculate the expected VCO free-running frequency (Equation 1). Compare it with the value found in Step
7, and record this value in Table 42–1. These two values should
agree within 10 percent. The loop's lock range is the difference,
$f_2 - f_4$, and is sometimes referred to as the *bandwidth*.

385

9. Determine the percent bandwidth for the 567 tone decoder using the following formula:

$$\% \text{ bandwidth} = \frac{f_2 - f_4}{f_0} \times 100$$

Record your result in terms of frequency and percent in Table 42–1. For the 567 tone decoder, the percent bandwidth is typically 14% if the input signal is greater than 200 mV rms.

10. The loop's capture range is the difference $f_3 - f_1$ and is never greater than the lock range. From your data, compute the capture range, and record this value in Table 42–1.

11. Starting with an input frequency of 200 Hz, slowly increase the input frequency until the LED goes out, which is the frequency you measured in Step 3 (f_1). During this time, you should notice that the VCO frequency (pin 5) remains constant at its free-running frequency, which you determined in Step 7. While the LED is lit, the loop is not locked and the VCO runs at its free-runing frequency.

12. Continue to increase the input frequency while the LED is unlit (the loop is locked). Observe that the output frequency now equals the input frequency over the lock range.

13. As an optional exercise, change the resistor between pins 5 and 6 to a different value, for example, 10 kΩ, and repeat the experiment. You should be able to determine the VCO free-running frequency and the lock and capture ranges.

WHAT YOU HAVE DONE

This experiment demonstrated the operation of a phase-locked loop using a 567 phase-locked loop tone decoder. The circuit uses a LED to indicate whether or not the phase-locked loop system is locked. The free-running frequency, lock range, capture range, and percent bandwidth characteristics were measured.

Name ———————————————— Date ———————

THE 567 PHASE-LOCKED LOOP TONE DECODER

OBJECTIVES/PURPOSE:

SCHEMATIC DIAGRAM:

Name _____ Date _____

DATA FOR EXPERIMENT 42

TABLE 42–1

f_1	Hz
f_2	Hz
f_3	Hz
f_4	Hz
Measured f_0	Hz
Expected f_0	Hz
% Bandwidth	%
Lock Range	Hz
Capture range	Hz

Name _____   Date _____

RESULTS AND CONCLUSIONS:

REVIEW QUESTIONS FOR EXPERIMENT 42

1. For the component values shown in Figure 42–1, the VCO free-running frequency is approximately
   (a) 200 Hz   (b) 400 Hz   (c) 600 Hz   (d) 800 Hz   (   )
2. When the 567 phase-locked loop tone decoder is locked, the LED is
   (a) unlit   (b) lit   (c) flashing   (   )
3. If the LED is lit and the input frequency is 200 Hz, the output frequency of the VCO is approximately
   (a) 200 Hz   (b) 400 Hz   (c) 500 Hz   (d) 700 Hz   (   )
4. If the LED is unlit and the input frequency is 500 Hz, the output frequency of the VCO is approximately
   (a) 300 Hz   (b) 400 Hz   (c) 500 Hz   (d) 600 Hz   (   )
5. The phase-locked loop's capture range is
   (a) always greater than the lock range
   (b) always less than the lock range
   (c) always equal to the lock range   (   )

NOTES

43

THE INTEGRATED-CIRCUIT
VOLTAGE REGULATOR

PURPOSE AND BACKGROUND

The purpose of this experiment is to demonstrate the operation of an integrated circuit (IC) voltage regulator, both as a simple, fixed voltage regulator and as an adjustable output voltage regulator.

In Experiment 7 voltage regulation is demonstrated using a zener diode. However, zener diodes in general do not have the ability to handle large load currents. On the other hand, IC regulators offer fixed output voltages with typical load regulation of less than 1 percent, internal thermal overload protection, as well as short-circuit protection.

Text References: 18–1, Voltage Regulation; 18–5, Integrated Circuit Voltage Regulators; 18–6, Applications of IC Regulators.

REQUIRED PARTS AND EQUIPMENT

Resistors (1/2 W):
- [ ] 47 Ω
- [ ] Two 100 Ω
- [ ] 150 Ω
- [ ] 220 Ω

- [ ] 1-μF capacitor
- [ ] 7805 +5 V voltage regulator
- [ ] 0–15 V dc power supply
- [ ] VOM or DMM (preferred)
- [ ] Breadboarding socket

USEFUL FORMULA

Adjustable-regulator output voltage

$$V_{OUT} = V_{REG} + R_2\left(I_Q + \frac{V_{REG}}{R_1}\right)$$

where $I_Q = 7$ mA   (typical)

PROCEDURE

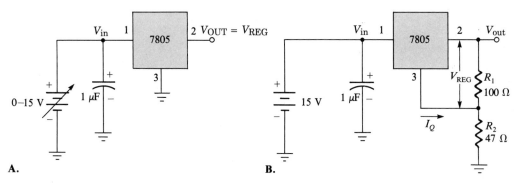

A.   **B.**

FIGURE 43–1 *Schematic diagram of circuits.*

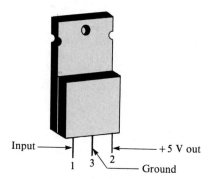

FIGURE 43–2 *Pin diagram of 7805 voltage regulator.*

1. Wire the circuit of the fixed output voltage regulators shown in the schematic diagram of Figure 43–1A. Connect the 1-μF capacitor as closely as possible to the 7805 regulator, as it greatly improves the transient response and stability of the internal circuitry of the 7805 regulator.
2. Apply power to the breadboard, and connect a VOM or DMM across the output of the regulator. Adjust the dc input voltage to the regulator, and record your result for each of the input voltages listed in Table 43–1. Plot the measured output voltage

versus the input voltage on the graph provided for this purpose. Such a plot of output voltage versus input voltage of a voltage regulator is known as its *dropout characteristic curve*.

From your measurements, observe that the output voltage of the regulator remains constant when the input voltage exceeds a given level. When this experiment is performed using a 7805 regulator, the dc output voltage remains constant at +5 V whenever the dc input voltage is greater than +7 V. The difference between the unregulated input and regulated output voltages of the regulator, termed the *output-input voltage differential,* is typically 2 V. On the other hand, the *dropout voltage* is that voltage below which the regulator circuit stops regulation. In general, this value is the sum of the regulated voltage and the output-input voltage differential. For the 7805 +5-V regulator, therefore, the dropout voltage is typically +7 V. If a type 7812 (+12 V) regulator were used, the expected dropout voltage would be +14 V.

3. Disconnect the power from the breadboard, and wire the circuit of the adjustable output voltage regulator shown in Figure 43–1B. For this part, you should use 1/2-W resistors.

4. Apply power to the breadboard, and set the dc input voltage to the regulator at +15 V. Then measure the dc output voltage using a VOM or DMM. From the equation given in the "Useful Formulas" section of this experiment, calculate the output voltage that you would expect to obtain. The typical quiescent current for the 7805 regulator is 7 mA. Record both the expected and the measured output voltages V_{OUT} in Table 43–2.

5. Change the value of R_2 according to the values listed in Table 43–2. Disconnect the power from the breadboard each time you change resistors. As in Step 4, record both the expected and the measured output voltages for each resistance value in Table 43–2. What happens to the output when resistor R_2 is 220 Ω?

You should find that the regulator output voltage is approximately +15 V, which equals its dc input voltage. It should be obvious that it is impossible to obtain an output voltage greater than the input voltage to the regulator.

WHAT YOU HAVE DONE

This experiment demonstrated operation characteristics of an 7805 (+5 V) integrated-circuit voltage regulator. The IC regulator was wired both as a simple fixed output voltage regulator as well as an adjustable voltage regulator. As long as the input voltage remained approximately 2 V above its rated dc output, the output voltage remained constant. In addition, the voltage regulator's dropout characteristic curve was graphed from measured data.

Name ______________________ Date ____________

THE INTEGRATED-CIRCUIT VOLTAGE REGULATOR

OBJECTIVES/PURPOSE:

SCHEMATIC DIAGRAM:

Name _____   Date _____

DATA FOR EXPERIMENT 43

TABLE 43–1 *7805 regulator Dropout characteristics.*

Input voltage	Measured Output Voltage
1 V	
2 V	
3 V	
4 V	
5 V	
6 V	
7 V	
8 V	
9 V	
10 V	
11 V	
12 V	
13 V	
14 V	
15 V	

Name _____   Date _____

TABLE 43–2 *Adjustable output voltage regulator.*

	$R_1 = 100\ \Omega$	$V_{IN} = 15\ V$	
R_2	Measured V_{OUT}	Expected V_{OUT}	% Error
47 Ω			
100 Ω			
150 Ω			
220 Ω			

NOTES

Name _____ Date _____

DATA FOR EXPERIMENT 43

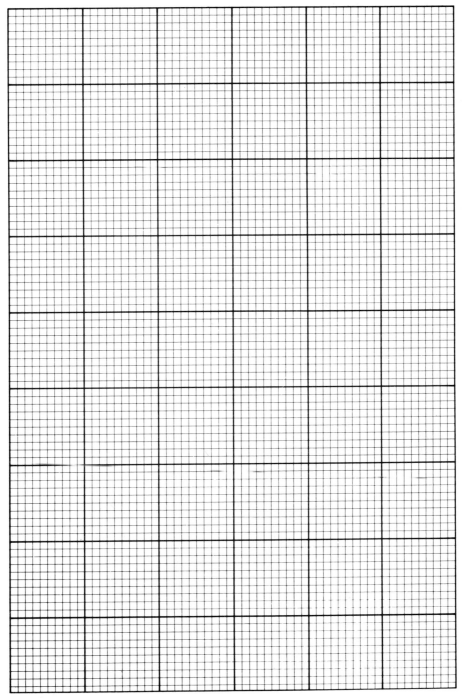

Name _____ Date _____

RESULTS AND CONCLUSIONS:

REVIEW QUESTIONS FOR EXPERIMENT 43

1. For the 7805 (+5 V) regulator in the circuit of Figure 43–1A, the output voltage remains constant for input voltages greater than approximately
   (a) +1 V   **(b)** +3 V   **(c)** +5 V   **(d)** +7 V   ( )

2. Using a 7805 (+5 V) regulator in the circuit of Figure 38–1A, if the dc input voltage is less than +5 V, the output voltage is
   (a) 0 V
   (b) +5 V
   (c) equal to the dc input voltage
   (d) an oscillating sine wave   ( )

3. For various load resistances, the output voltage of the regulator circuit of Figure 43–1A
   (a) stays essentially constant
   (b) increases with increasing load resistance
   (c) decreases with increasing load resistance
   (d) oscillates   ( )

4. Using the adjustable regulator circuit of Figure 43–1B, if the dc input voltage is +15 V, the quiescent regulator current is 7 mA, $R_1 = 100\ \Omega$, and $R_2 = 100\ \Omega$, the the output voltage is approximately
   (a) +5 V   **(b)** +8 V   **(c)** +11 V   **(d)** +15 V   ( )

5. If R_2 is changed to 270 Ω in Question 4, the output voltage of the regulator is approximately
   (a) +5 V   **(b)** +10 V   **(c)** +15 V   **(d)** +20 V   ( )

400

REQUIRED PARTS AND EQUIPMENT FOR THE EXPERIMENTS

TABLE A–1 *Resistors (1/4 W minimum; all may be 1/2 W).*

Quantity	Value	Radio Shack P/N
1	10 Ω	271-1301
1	68 Ω	—
2	100 Ω	271-1311
1	150 Ω	271-1312
1	180 Ω	—
1	220 Ω	271-1313
1	330 Ω	271-1315
1	470 Ω	271-1317
1	560 Ω	—
1	680 Ω	—
4	1 kΩ	271-1321
1	1.5 kΩ	—
1	2.2 kΩ	271-1325
2	2.7 kΩ	—
1	3.3 kΩ	271-1328
2	3.9 kΩ	—
1	4.7 kΩ	271-1330
3	6.8 kΩ	271-1333
2	10 kΩ	271-1335
2	12 kΩ	—
1	15 kΩ	271-1337
1	22 kΩ	271-1339
1	27 kΩ	271-1340
1	47 kΩ	271-1342
1	68 kΩ	271-1345
2	100 kΩ	271-1347
1	180 kΩ	—
1	560 kΩ	—

TABLE A–2 *Resistors (1/2 W).*

Quantity	Value	Radio Shack P/N
1	47 Ω	271-009
2	100 Ω	271-012
1	150 Ω	271-013
2	220 Ω	271-015
1	1 kΩ	271-023
1	2.2 kΩ	271-027
1	100 kΩ	—
1	1 MΩ	—

TABLE A–3 *Potentiometers (Single turn).*

Quantity	Value	Radio Shack P/N
1	5 kΩ	271-217
1	10 kΩ	271-218
1	100 kΩ	271-220
1	1 MΩ	271-229

TABLE A–4 *Capacitors.*

Quantity	Value	Radio Shack P/N
1	0.0022 μF	—
2	0.0047 μF	—
2	0.01 μF	271-1051
2	0.033 μF	—
2	0.1 μF	271-1053
1	1 μF electrolytic	272-1419
2	2.2 μF electrolytic	272-1420
1	4.7 μF electrolytic	272-1012
1	10 μF electrolytic	272-1013
2	100 μF electrolytic	272-1016
1	470 μF electrolytic	272-1018

TABLE A–5 *Solid state devices.*

Quantity	Description	Radio Shack P/N
2	1N914 diode	276-1620
4	1N4001 diode, 50 PIV	276-1101
1	1N735, 6.2 V, 400 mW zener diode or 1N4735 (1 W)	276-561
1	LED (red)	276-041A
1	2N2646 UJT (or HEP 310)	276-2029
2	2N3904 npn transistor	276-2016
1	2N3906 pnp transistor	276-2034
1	MPF102 n-channel FET	276-2062
1	40673 n-channel depletion-mode MOSFET, or equivalent	—
1	NTE 465 n-channel enhancement-mode MOSFET, or equivalent	—
1	VK 67AK n-channel enhancement-mode VMOSFET, or equivalent	—
1	200-V, 6-A SCR	276-1067
1	555 timer (8-pin mini-DIP)	276-1723
1	567 tone decoder	276-1721
2	741 op-amp (8-pin mini-DIP)	276-007
1	7404 TTL hex inverter	276-1802
1	7442 TTL 1-of-10 decoder	—
1	7474 TTL dual D-type flip-flop	276-1818
1	7490 TTL decade counter	276-1808
1	7805 +5 V regulator (TO-220 pkg)	276-1770

TABLE A–6 *Miscellaneous.*

Quantity	Description	Radio Shack P/N
1	12.6 VCT transformer	273-1505
1	Breadboarding socket	276-174
2	VOM, 50 kΩ/V minimum or digital multimeter	28-4014
1	Dual trace oscilloscope	—
1	Signal Generator	—
2	0–15 V dc power supplies	—
1	Frequency counter	—
1	12 V relay (SPDT or DPDT)	—
1	SPDT switch	—